CISM COURSES AND LECTURES

Series Editors:

The Rectors of CISM
Sandor Kaliszky - Budapest
Mahir Sayir - Zurich
Wilhelm Schneider - Wien

The Secretary General of CISM
Giovanni Bianchi - Milan

Executive Editor
Carlo Tasso - Udine

The series presents lecture notes, monographs, edited works and
proceedings in the field of Mechanics, Engineering, Computer Science
and Applied Mathematics.
Purpose of the series is to make known in the international scientific
and technical community results obtained in some of the activities
organized by CISM, the International Centre for Mechanical Sciences.

INTERNATIONAL CENTRE FOR MECHANICAL SCIENCES

COURSES AND LECTURES - No. 408

COMPUTATIONAL INTELLIGENCE IN DATA MINING

EDITED BY

GIACOMO DELLA RICCIA
UNIVERSITY OF UDINE

RUDOLF KRUSE
OTTO-VON-GUERICKE UNIVERSITY

HANZ-J. LENZ
FREE UNIVERSITY BERLIN

 Springer-Verlag Wien GmbH

This volume contains 35 illustrations

SPIN 10768082

In order to make this volume available as economically and as
rapidly as possible the authors' typescripts have been
reproduced in their original forms. This method unfortunately
has its typographical limitations but it is hoped that they in no
way distract the reader.

ISBN 978-3-211-83326-1   ISBN 978-3-7091-2588-5 (eBook)
DOI 10.1007/978-3-7091-2588-5

PREFACE

This volume contains papers accepted for final publication and read at the 3rd workshop on "Computational Intelligence in Data Mining", Udine (Italy), Oct 8-10, 1998.

As its preceeding ones, this international conference was conveyed by the International School for the Synthesis of Expert Knowledge (ISSEK) and held in Palazzo del Torso of the Centre International des Sciences Mécaniques (CISM), Udine.

The workshop was organised jointly by Prof. G. Della Riccia (University of Udine), Prof. R. Kruse (University of Magdeburg), and Prof. H.-J. Lenz (Free University Berlin). The talks were given by researchers who were recruited from various fields and individually invited. The various backgrounds of the participants were intended, and improved the quality of the round-table discussions and, hopefully, stimulated further studies.

Computational Intelligence and Data Mining are hot topics of current research. The extension of their semantics is still under discussion.

Computational Intelligence incorporates for example techniques like data fusion, uncertain reasoning, heuristic search, learning, and soft computing. Data Mining focuses on unscrambling unknown patterns or data structures in very large data sets. In this sense it can be considered as explorative data analysis on massive data. Evidently, it makes sense to join both areas.

In section 1: "Discovering Structures in Large Databases" A. Siebes aims for a unified view in his paper 'Data Mining and Statistics - A System Point of View'. W. Kloesgen gives an outlook on 'Subgroup Mining', and C. Borgelt, J. Gebhardt and R. Kruse present 'Data Mining with Possibilistic Graphical Models'.

In section 2: "Data Fusion and Possibilistic or Fuzzy Data Analysis" S. Benferat's paper is devoted to 'An overview of possibilistic logic and its applications to nonmonotonic reasoning and data fusion'.

H.-J. Lenz and R. Müller analyze the coherence problem between data and a corresponding nonlinear model in 'On the Solution of Simultaneous Fuzzy Equation Systems', and M. Berthold contributes to the field of outlier detection by 'Learning Fuzzy Models and Potential Outliers'.

Section 3 is entitled "Classification and Decomposition". The contribution of F. Schwenker, H.A. Kestler and H.G. Palm on 'An Algorithm for Adaptive Clustering and Visualisation of High Dimensional Data Sets' combines adaptivity and cluster techniques. The extended abstract 'Computing with Decomposable Models' of F.M. Malvestuto will be of interest for readers familiar either with graphical models or database modelling.

Finally, in section 4: "Learning and Data Fusion" H.-D. Burkhard studies learning problems of special multi-agents systems in 'Learning in Computer Soccer' and E. Rödel and H.-J. Lenz treat consistency problems of fused data in a paper entitled 'Controlling Based on Stochastic Models'.

According to the great success and echo of all those ISSEK workshops, we are glad to announce the next workshop on "Fusion and Perception" for October 2000.

The editors of this volume thank very much our authors for submitting their papers almost in time, and Mrs. Angelika Wnuk, Free University Berlin, for her superb 'back stage' and secretarial work throughout all the phases of the workshop.

We would like to thank the following institutions for substantial help on various levels:

- *The International School for the Synthesis of Expert Knowledge (ISSEK) again for promoting the workshop.*
- *The University of Udine for administrative support.*
- *The Centre International des Sciences Mécaniques (CISM) for hosting a group of enthusiastic scientists.*

On behalf all participants we express our great gratitude to CASSA di RISPARMIO di UDINE e PORDENONE and also to FONDAZIONE CASSA di RISPARMIO di UDINE e PORDENONE for their financial support.

Giacomo Della Riccia
Rudolf Kruse
Hans-J. Lenz

CONTENTS

Page

DATA MINING AND STATISTICS
A SYSTEMS POINT OF VIEW

A. Siebes
CWI, Amsterdam, The Netherlands

1 Introduction

Moore's law has never been so obvious as it is now. New PC's are equiped with hundreds of Megabytes of main memory, many Gigabytes of secondary storage and processors approaching a Gigaherz clockspeed. Fortunately[1] the need for such resources is growing just as fast if not faster.

On the storage side there is a version of Moore's law that states that the amount of data stored triples every two years. Supermarkets, Banks, Telephone Companies, and many others now routinely store all transactions their clients make. This is partly because of mundane reasons such as accounting, billing, and supply management. Other reasons can be found in the ever more competitive economy. Since stock costs money, supermarkets prefer just in time delivery. This means that they should be able to predict what they will sell tomorrow; to support this, a database registers the sales of today. In the same vein, longer term trends should also be recognized. If low-fat food becomes popular, the image of the supermarket should reflect this trend before it is too obvious. Similar considerations hold for most companies.

In order to understand the behaviour of clients and predict what they'll do in the (near) future, *models* are necessary. Models that give this insight and allow accurate predictions. In other words, the enormous amounts of data that are collected should be refined into such models.

Luckily, the growth in computer resources allows the analysis of huge amounts of data. Moreover, there is a rapid growth in tools and techniques for data analysis. Not only in its traditional habitat, i.e., Statistics, but also in Computer Science. The last decade has given rise to the quickly growing area called *Data Mining*. This area combines techniques from Statistics, AI, Databases, Visualization, and many others to support the (semi-) automatic discovery of models from (sometimes huge) data sets.

True to the spirit of the Udine workshops on AI and Statistics, this paper gives both an introduction to Data Mining as well as a survey of one research direction. That is, it gives an introduction to data mining from the systems point of view: "how does one built a system for data mining?". Or, more in particular, what is the KESO architecture and what was its rationale?

---

[1] Or, unfortunately if *you* have to fork out the money for new machines all the time

It *does not* present a survey of the complete area. Many topics are only metioned; if at all. As far as algorithms are concerned, we only discuss a few and many algorithm classes are ignored all together. The resaon for this is firstly that as a far as the KESO architecture is concerned, the algorithms are only meant as an illustration. Secondly, because a detailed discussion of mining algorithms would need a book rather than an article.

Readers that are interested in a broader and more detailed survey are refered to books like [16]. Much of the current research can be found in the proceedings of the KDD conferences [17, 45, 23, 2, 7], the PKDD conferences, [28, 52, 51], the PAKDD conferences, [33, 49, 50], the IDA conferences [32, 21], journals like *Machine Learning* and the *Data Mining and Knowledge Discovery* journal. Finally, the website at kdnuggets.com gives a good entry point to the web.

The first section of this article is the introduction you have now almost finished reading. The next section discusses an informal definition of Data Mining. Moreover, it presents a brief description of the KDD process of which Data Mining is just one, albeit important, step. Next, we'll discuss Data Mining algorithms. Starting with algorithms for two well known models, viz., *association rules* and *classification trees*, an abstract form of mining algorithms is introduced. In the next section it is shown how the structure of such abstract mining algorithms is faithfully reflected in the architecture of the KESO system. Following this architectural discussion it is shown how mining algorithms fit in the framework. Next to our example algorithms, viz., association rules and decision trees, these are algorithms for the discovery of interesting subgroups and two statistical algorithms, viz., for regression and for projection pursuit. In the next section the architectural discussion is continued with a discussion on how database techniques can be used to speed up the mining process. The final section presents the conclusions of this paper.

2   Data Mining and Knowledge Discovery in Databases

2.1   Data Mining

In the introduction, Data Mining has been introduced as an area that combines techniques from Statistics, AI, Databases, Visualization, and many others to support the (semi-) automatic discovery of models from (sometimes huge) data sets. Although this is very much true, it doesn't help too much in understanding what data mining is. For this I very much like the following pseudo definition:

the induction of understandable models and patterns from a database

In other words in Data Mining we have initially a large (possibly infinite) collection of possible models (patterns) and a (finite) database. Data Mining should result in those models that describe the database best (those patterns that fit (part of) the database best).

I like this definition so much, because it highlights some of the most important problems associated with Data Mining, viz., *models* and *patterns*, *induction*, and *databases*. The remainder of this subsection is devoted to these aspects.

2.1.1   Models and patterns

As discussed above, Data Mining is the induction of understandable models and patterns. There are two immediate questions. Firstly, what is the difference between models and patterns? Secondly, what kind of models and patterns are induced?

For the first question, one should realize that traditionaly statisticians have been concerned with models that describe the whole dataset. In data mining, one is often only interested in models that describe only part of the database. For example, one is after a description of the best potential customers. All other potential customers are simply not interesting. Phrased differently, patterns are partial models. Now that the distinction is understood, I'll use *model* as the generic word; if only to keep the paper readable (and writable).

Answering the second question on a rather abstract level is relatively easy. The model one is after can always be seen as a function (or, even more general, as a relation).

On a more concrete level, it is probably impossible to give a comprehensive answer. The kind of model one tries to induce depends on at least on the question one tries to solve and on what kind of data one has in the database. So, rather than aiming at a comprehensive answer I list some classes of models that have receiced much attention in the Data Mining community; the reader is urged to consult e.g., [16] to get a more detailed picture.

Functional Models when one of the attributes is a (real-valued) function of the other attributes, the aim is to discover this function. A popular type of models in this class are *Neural Networks*; see, e.g., [24] for a nice introduction.

Dependencies where one tries to discover significant dependencies between attributes. Examples are Association Rules ([3]) and, of course, functional dependencies ([34]). An example of an association rule is, e.g., the discovery that if a customer buys diapers he will often buy a sixpack. More general, it are expressions of the form:

$$X \to Y$$

which state that clients that bought the products in X have a high(er) probablity of bying also the products in Y. We will discuss association rules in more detail later in this paper.

Classification learning a function that maps tuples into one of several pre-defined classes. There are many different types of models for this class, a popular one are so-called classification (or decision) trees such as in Figure 1.

Classification trees will be discussed in more detail later in this paper.

Clustering which is somewhat similar to classification, only the classes are not pre-defined. It is both an interesting and a hard Data Mining problem.

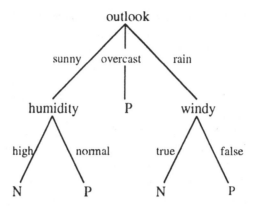

Classification trees will be discussed in more detail later in this paper.

As already stated above, what kind of model class one uses depends on the kind of *mining question* one tries to answer as well as on the kind of data one has. For example, classification models are a special type of functions, and so are functional dependencies. But function expressions are not always the best choice.

Discovering Models   When one has decided on what kind of model one is after, the best (relative to the database) model has to be discovered. There are many kinds of techniques. In fact, the reader should note that model induction is not only a problem studied in Computer Science, but has also received much attention in Statistics. Many of the techniques that can be found under such headings as *Exploratory Data Analysis* [48] and *Multivariate Analysis* [35] can be seen as a kind of data mining algorithms. Especialy modern techniques in these areas, such as *Projection Pursuit* [18, 26] (see also later in this paper) are very close in spirit to current Data Mining research.

What technique should be used depends on what kind of model one is after, the type of data in the database, and the domain knowledge one has on the problem at hand. For example, if the model should be a function, the data is all numerical and one knows the function expression except some parameters one should use *regression*. However, if one does not know the function expression, techniques such as Neural Networks [24] and Genetic Programming [29, 30] can be used. Similarly, for cluster models, techniques range from simple cluster analysis techniques [12] to Bayesian classification [8].

In some cases, such as regression, there are techniques that allow one to compute the best (defined in a formal way) model directly. In others, such as association rules, one can effectively enumerate all possibilities and choose those that are good enough (as defined by the user). In still other cases, there are no known direct computations and an exhaustive search is prohibitively expensive.

In such latter cases *heuristics* are used that yield a reasonable (not necessarily best) model. Examples in this class are (again) Neural Networks and Genetic Programming.

Quite often there is more than one technique available for the problem at hand. One of the aims of the ESPRIT-III project STATLOG was to find out which algorithms performed best for some of these problems, depending on, a.o., database characteristics. The result in this respect was: there are no clear cut winners, see [36]. In other words, the data miner should try some different techniques and pick the best result.

2.1.2 Induction

Induction as an inference scheme means that we generalise from a *finite* number of facts (the database). The finite basis of facts on which we base our conclusions leads to the *problem of induction:* there is no absolute guarantee of correctnes. For example, before the discovery of Australia, Europeans had seen millions of swans, all of them white. The generalisation *Swans are white* seemed more than reasonable. However, Australia provided examples of *black* swans.

Moreover, given a finite number of facts there are, in general, infinitely models that describe the database exactly. However, such absolutely correct models often perform badly on new facts. This problem is often called *overfitting*. Since the whole point of Data Mining is often to use the induced models to predict in new situations this is bad news. It means that good fits are not exact fits.

In other words, there is no guarantee that induction yields the *true* model. In fact, one should not care too much whether or not a *true* model exists at all. One should aim for a good model, that is a model that can be used for the purposes it was designed for with reasonable confidence. That is, one should use a *quality measure* that guides to such good models.

Clearly, Data Mining is not the first human endavour that encounters these problems. In fact, one could say that the very strict methodology used by main stream (or frequentist) statisticians ([46, 47]) is mainly a safeguard against the dangers of induction. A somewhat more relaxed approach is taken by Bayesian statisticians ([37]) and computer scientists who use the MDL principle [20] or, more general, Kolmogorow complexity [31].

Data miners approach these problems pragmatically. On the one hand one can take statistical safety measures against overfitting, as well as methods inherited from Bayesian statistics and MDL. Moreover, it is common practice to leave some data out of the induction process to assess the quality of the model on unseen data. On the other hand, there is "the Human in the loop". That is, at the end of the day the data miner decides how good a result is. This pragmatic approach is, by the way, is one of the prime motivations for insisting on understandable results.

2.1.3   The Database

The problem of induction is worsened by the fact that we use databases rather than specially collected data. The first problem with databases is that they invariably contain noisy and/or contradictory data. While data cleaning can alleviate this problem to a certain extend, this is often too costly or downright impossible. This means that data mining algorithms have to be robust against errors in the data.

A second class of problems is caused by the fact that most databases are maintained for other reasons than data mining. In the words of David Hand, data mining is *secondary analysis*. The first such problem is that databases are not very often random samples from a population. For example, banks only have credit histories of those clients that they gave credit. In other words, a credit model induced from the credit database cannot be used automatically for those clients one would reject under the old credit-regime. This problem is called *reject inference*.

Reject inference is a problem of "tuples missing in the database", similarly, there are quite often attributes missing in the database. Such attributes are missing because they are not important for the original goals of the database, or because it is too costly or simply forbidden to collect and maintain them. That means that information that could of vital importance for the data mining task at hand is simply not present. Thus, even if absolute models could exist in theory, in Data Mining one often finds statistical (or probabilistic) models.

Again, the data mining community approaches these problems pragmatically. To a large extend the problems simply mean that one should not hope for extremely good models. And, of course, if the models discovered with data mining are better than the current practice, they save money.

2.2   The KDD process

Now that it is understood what data mining is, it is time to discuss how data mining is done in practice. Given the pitfalls described above, it should be clear this is a process that has to be taken with care. Such a procedure is called *Knowledge Discovery in Databases*. An attempt to standardize this process has been undertaken in the ESPRIT project CRISP. I'll first describe the CRISP model and then I'll discuss some other apsects of data mining in practice.

2.2.1   The CRISP model

Unfortunately, the CRISP project hasn't used the now standard distinction between data mining and the KDD-process of which data mining is one step. The process model is therefore christened the CRISP-DM model. The reader who is interested in more details than I can give here is referred to the website at crisp-dm.org where he can find the current description of the process model.

The CRISP-DM description is in terms of a hierarchical process model consisting of four levels of abstraction: *phases, generic tasks, specialized tasks*, and *process instances*.

At the top level, the data mining process is organized into several phases, just as in earlier descriptions of the KDD-process; see e.g., [15]. Each phase consists of several second-level generic tasks, i.e., the tasks that occur in all mining projects. The third level, the specialized task level, describes how actions in the generic tasks should be carried out in certain specific situations. The fourth level, the process instance level, is a record of actions, decisions, and results of an actual data mining project.

One should realize that the description of phases and tasks as discrete steps performed in a specific order represents an idealized sequence of events. In practice, many of the tasks will be performed in a different order and it will often be necessary to backtrack to previous tasks and repeat certain actions.

It would take us too far to describe all levels in detail, but a brief look at the first two levels should illustrate the main idea. The phases with their generic tasks are as follows:

Business Understanding: before one can start any analysis, one has to understand what business problem is to be solved, what are the objectives and when is a result considered a success. Moreover, one should in this phase evaluate the situation, e.g., what are the resources and the requirements and what are the risks? Finally, the business problem should be translated into a mining problem which states both the goals and the success criteria,

Data Understanding: once the problem is clear, one should collect the available data and explore this data, e.g., to assess the quality of the data but also to get some first insight.

Data Preparation: most, if not all, data mining systems assume that all the data is in one table. The main goal of this phase is to construct this table. Tasks in this construction are: selecting data, creating derived attributes, cleaning the data, and, of course, construction of the actual table.

Data mining: once the table is created, the actual data mining can take place. Common tasks are the selection of a mining technique, create a test design, perform the mining, and, test the mining results using the test design.

Evaluation: now that the models are found and tested statistically, it is time to evaluate the results with respect to the success criteria. That is, do the models help in solving the original problem?

Deployment: once adequate models are found, they should be deployed in the company to actually solve the original business problem.

2.2.2   Mining in Practice

Data mining always begins with a business problem. That is, some business problem is translated into a data mining problem. For example, the business problem may be that one wants maximal return of investment on a mailing for a given, fixed budget. This may be translated to the data mining problem: "find me those customers that are likely to order the item on offer".

After solving the mining question, one has to return to the original business question. That is, one has to make the step from acquired knowledge to decision making. Unfortunately, this is often less easy than assumed. The reason is that in data mining, one often has to use the data that is available and that is not always the data one would have liked to have.

Take the marketing example introduced above. In the ideal case, one first sends this mailing to a random subset of the population and analyzes the behaviour of this subset. Subsequently, one uses the results of this analysis to determine whom to send the mailing to. If all of this has been done carefully, one can predict the return on investment of the final mailing pretty accurately. In fact, that is exactly what a "random subset of the population" means.

There are many reasons why the ideal case is not often the real case. Sometimes there are not enough resources (time, money or otherwise) to produce and analyze such a proper random subset of the population. In other cases, it is simply impossible to acquire the required input data. In all such cases, one has to do with the data that is available. For example, rather than the data of a trial mailing, one has to do with the data on what clients have what products.

The Ideal Case   Even in the ideal case, i.e., the available data matched the analysis requirements some care has to be taken in applying the results of data mining in practice. An important class of problems is caused by the costs of wrong decisions. In simple cases, such as sending mailings, these costs are limited and well known. In other cases, e.g., in a fraud examination such costs are almost boundless and far harder to predict. What happens for example, if you accuse one of your largest customers wrongly of fraud?

In those cases where the potential costs and profits can be reasonably estimated beforehand, they should be taken into account even before the actual mining takes place. For many popular mining tasks, such as classification, there are techniques available that use such cost matrices.

The theoretical framework in which this fits is called *decision theory*, quite often this entails to Bayesian decision theory; see, e.g., [43]. As usual in the Bayesian approach, the statistical analysis yields a probability distribution over all possible models. Combining this with the costs, one can compute the estimated costs of each possible decision and take the one that minimizes these costs, or, phrased more positively, maximizes the profits. More formally, the Bayes optimal prediction for a tuple x is the class that minimizes the conditional risk:

$$R(i|x) = \sum_j P(j|x)C(i,j)$$

in which $C(i,j)$ are the costs of predicting class i for x whereas it should have been j and $P(j|x)$ is the probability that x belongs to class j.

Even if the costs are hard to estimate, the Bayesian framework in principle tells you what to do, viz., make a probablity distribution over the possible costs.

Although this makes for an elegant theory, it is not always that applicable in practice. Especially in cases where not too many clients will be singled out, such as for fraud, it makes sense to use common sense. That is, to have a human expert peruse the results and ultimately decide which cases do warrant a follow-up and in what way.

As already said above, if known beforehand, cost matrices should be used as early in the mining as possible. Either by using a tailor made algorithm that is cost-sensitive, or by using a procedure that converts error-based classifiers into cost-sensitive classifiers. One approach in this latter direction is by using stratified sampling. That is, by changing the frequency of classes in the data based on their costs; see [6, 38, 13].

Reality Strikes As argued above, quite often one has to cope with the problems of using data that does not realy match the analysis requirements. In a sense, this can be seen as falling under the heading of missing (and superfluous) information. It is a severe case, however, in that complete attributes can be missing as well as (very informative) tuples.

A prime example of the latter occurs in *credit scoring*. If a bank wants to establish a new credit scoring function, i.e., whom should we give credit, they do not have a credit-history on those cases that they refused. In general, however, they do retain the application details of those loans that were refused. Making inferences on the probable behaviour of those refused creditors is what they call *reject inference* and is absolutely crucial in credit scoring.

The general approach towards this problem is, again, to take it into account during the mining as much as possible. There are different ways to attack this problem, most of them rely on model assumptions, see [11] for an overview.

The final, and probably most frequent, case to consider is that you know (and you ought to know) that your data doesn't really match your requirements and there isn't much you can do about it. In such cases, all you can do is use your common sense and start by distrusting your results.

Consider again our earlier example, you try to find the right groups to send a mailing to based on the products clients actually own. If you proceed by classifying clients in owners and non-owners of, say, product X, it has possible that you'll find groups, say, retired lawyers, that have a high percentage of X owners. Does this mean that retired lawyers do not yet own X have a high probability of reacting to your mailing? Not necessarily!

Perhaps your famous product X is a set of law books, why would these retired lawyers buy your books? They would rather play golf than read about the law.

How do you find out? Well, there are no golden rules that will ensure that you'll never make such mistakes. But, you can take precautions that will help you. The first rule is: mistrust your results. If you have used some tree-induction algorithm, you should be aware that these tend to prune off unsignificant branches. The problem is that significance is only used in the statistical sense. That is, the (statistically) insignificant bits may have had the information that would tell you why this group scores so high. So, check the insignificant bits. In other words, try to distinguish the owners from the non-owners in this group. Do this by any means and algorithms you have got.

The second, and most important rule, is that results only make sense if they make sense to a domain expert. That is, let a domain expert scrutinize the proposed groups. Does it make sense to her that you would like to send the maining to those people or not. If not, don't do it. Whatever promises the statistics of your model make.

3 Mining Algorithms

Now that we know what data mining is and have some feeling for how it should be approached in practice, it is time to look at mining algorithms. This section consists of two parts. In the first part two example mining algorithms are introduced. The goal of this part is not to present the state of the art algorithms for particular problems, but to give insight in how data mining algorithms actually work. This insight is exploited in the second part of this section where an abstract definition of data mining algorithms is presented.

3.1 Two Examples

In some cases, mining algorithms exist that find all good models guaranteed. In most cases, however, heuristic techniques have to be used and there is no guarantee that the discovered models are (near) optimal. In this subsection we give an example of both cases.

3.1.1 Association Rules

For association rules, we assume that we have a table r with schema $R = \{A_1, \ldots A_n\}$, in which A_i is a binary attribute (note, more general forms exist, see, e.g., [4]). In the prototypical example, the attributes denote the items for sale in a shop and the rows in the table are the shopping baskets presented at the check-out. That is, a row has a 1 for an attribute if that item was present in the basket and a 0 otherwise. Association rules can then be used to discover cross selling between products in the supermarket.

The general form of an association rule is as follows:

$$X \to Y \ (t_1, t_2)$$

in which X and Y are disjoint sets of attributes (sometimes called *items* in this context) and the parameters t_1 and t_2 describe the quality of the rule; t_1 is called the rule confidence and t_2 is the rule support. To explain these parameters, let $s(X)$ denote the number (or fraction) of tuples in the database that have a 1 for all attributes in X. Then, for the general rule given above, we have:

1. $s(XY)/s(X) = t_1$ and

2. $s(XY) = t_2$

In other words, if X occurs in a tuple, with probability t_1 Y also occurs in that tuple, and the X and Y co-occur in a fraction t_2 of the entire database.

The problem is given minimal thresholds for confidence th_1 and support th_2 find all association rules whose confidence and support exceed these minimal thresholds. The crucial observation for this algorithm is that we can generate these association rules in two steps:

1. find all sets X whose support exceeds the minimal threshold and their support. These sets are called frequent (or large) sets.

2. then test for all non-empty subsets Y of frequent sets X whether the rule holds with sufficient confidence.

For the generation of frequent sets, there is another important observation, viz., a set X can only be frequent if all its (non-empty) subsets are frequent. Let $C(i)$ denote the sets of i items that are potentially frequent and let $F(i)$ denote the frequent sets with i items. The observation then gives the following algorithm:

Find frequent sets
$C(1) := \{A | A \in A\}$
$i := 1$
While $C(i) \neq \emptyset$ **do**
  $F(i) := \emptyset$
  For each $X \in C(i)$ **do**
    If $s(X) \geq th_2$ **then** $F(i) := F(i) \cup \{X\}$
  $i := i + 1$
  $C(i) := \emptyset$
  For each $X \in F(i-1)$ **do**
    If $\exists Y \in F(i-1)$ that shares $i-2$ items with X **then**
      If All $Z \subset XY$ of $i-1$ items are frequent **then**
        $C(i) := C(i) \cup \{XY\}$

Now that we have the frequent sets, we can generate the association rules from them:

Generate Association Rules
  For each frequent set X **do**
    For all non-empty $Y \subset X$ **do**
      If $s(X)/s(X \setminus Y) \geq th_1$ **then**
        Output $X \setminus Y \to Y$

3.1.2   Classification and Classification Trees

For the classification problem, we assume that our database has n regular attributes, say $A_1, \ldots, A_n$. Moreover, there is an extra attribute A_{n+1} that assigns one of the pre-defined classes $C_1, \ldots, C_k$ to each of the tuples in the database.

The goal is to device a procedure that assigns the correct (or better, most likely) class to a *new* tuple $(v_1, \ldots, v_n)$.

There are various ways in which one can approach this problem. First of all, one could device a technique that compares the new tuple with the tuples in the database and picks the most likely class on the basis of these comparisons. *Naive Bayes* and *k-Nearest Neighbours* are examples of this approach [42]. For naive Bayes, one assumes that the attributes are conditional independently distributed given the class and computes the most likely class using Bayes rule and the marginal distributions found in the database. Although these assumptions may seem overly strict and not very often true in practice, it is well known that Naive Bayes performs very well in practice. One could say that Naive Bayes is not too naive.

For k-Nearest Neighbour one needs a distance function. Using this function one computes the k tuples in the database that are closest to the new tuple. The class C_j that occurs most frequently among these k nearest neighbours is the class that is assigned to the new tuple.

Another approach is to induce a model that allows one to assign a class to a new tuple without consulting the database. Within this approach, there are two sub-approaches possible. Either one induces what distinguishes one class from another or one induces what characterises each class. An example of the former approach is *Discriminant Analysis* and of the latter *Classification Trees* are a prime example.

Discriminant analysis is easiest understood if we assume that we just two (regular) real valued attributes and two classes. Then we simply plot the points in the database, label all points with their class and draw the line that distinguishes the two classes best. See [42] for more details. Classification trees are the topic of the remainder of this discussion.

As an aise, a final approach is to consider functions as models for classification problems. That is, induce a function that assigns the correct class to a tuple. In that approach, one could, e.g., employ Neural Networks. This option is discussed in depth in [5] and [42].

One of the most popular approaches to classification problems is classification by *trees*. These trees are often called *Decision Trees* in the Computer Science literature, whereas statisticians often call them *Classification Trees*. In this paper, I'll use the latter term. These trees are models of the form as in Figure 1.

This tree should be read as follows. If the outlook is sunny and the humidity is normal, we will play golf. However, if the outlook is sunny and the humidity is high, we won't play golf.

Clearly, such a model is far easier to understand than a Neural Network. The question now is, of course, how do we induce such a tree from a database? The number of possible trees prohibits an evaluation of all possible trees to pick out the best. For example, the largest possible tree would have all attributes on all paths from the root to a leaf and would have branches for all relevant attribute values at each node. The smallest possible tree would have no splits at all. And each tree in between is also possible. In other words, heuristics have

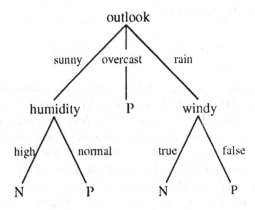

Figure 1: A Classification Tree for Golf

to be employed.

Most algorithms that build classification trees have the same simple underlying algorithm; they differ, of course, in the heuristics employed. In its description, I'll use $\mathcal{A}$ to denote the set of regular attributes $A_1, \ldots, A_n$.

1. Set-up the tree T with one node N that represents the complete database.

2. If node N covers one of the classes C_i *good enough*, lable that node with class C_i and stop.

3. Otherwise, *pick an attribute* $A \in \mathcal{A}$ and partition the set of objects covered by N on the values they take for A in the sets $N_1, \ldots, N_m$. Create a branch for each of the N_i.

4. Apply the procedure recursively to each of the child nodes.

The algorithms differ in what *good enough* means, what attribute A is chosen, and how the set of objects is partitioned. In ID3 [40] *good enough* is interpreted as saying "all nodes covered by N belong to the same class", the choice of attribute A is explained below and N is partitioned using all distinct values used for attribute A.

For the choice of attribute A, assume that node N covers a subset S of the database. If we denote by $P(C_i, S)$ the fraction of tuples in S that belong to class C_i, the class-entropy of N is given by:

$$Ent(N) = -\sum_{i=1}^{i=k} P(C_i, S) log(P(C_i, S))$$

If we split N on the attribute values taken for attribute A, we get, say, nodes $N_1, \cdots, N_r$ which cover the partitioning $S_1, \cdots S_r$ of S. The resulting class-entropy after this split is given by:

$$E(N, A) = -\sum_{j=1}^{r} \frac{|S_j|}{|S|} Ent(S_j)$$

The *quality* of this split, called the *information gain* is defined as *Gain(N, A)* = *Ent(N)* - *E(N, A)*. ID3 makes the split that gives the greatest gain.

It should be clear that ID3 is but one of the (many) possibilities to flesh out the skeletom algorithm presented in the previous subsection. First, one could use different criteria to decide which attribute is taken next. Quinlan used a different *Gain-criterium* for the C4 family of algorithms [41].

In CART [6], the choice of the splitting attribute is based on *impurity measures*. Such an impurity measure quantifies how the tuples covered by a node are distributed over the classes. Using the impurity measure, gains are computed using, e.g., an information theoretic approach (like ID3) or the Gini index.

Similarly, one could decide not to split a node with all possible attribute values. Especially in the case of continuous attributes this leads to very fat trees. For example, both C4.5 [41] and CART [6] handle such cases differently. Moreover, both CART [6] and GID3* [14] don't necessarily split discrete attributes on all values.

Finally, I should note that growing the tree is but on aspect of tree induction. The second step is to *prune* the resulting tree, e.g., to avoid overfitting. The reader should consult, e.g., [6] for more details.

3.2   The General Form

One of the main challenges in building a data mining system is the flexibility necessary both to support the current variety of algorithms, such as the two introduced above, and to extend it easily with new kinds of data mining algorithms. In the KESO project this challenge was met by basing the system on the common structure underlying most, if not all, current data mining algorithms. This structure is the topic of this subsection.

Without loss of generality ([1]) we assume that our database consists of one *Universal* relation DB, with schema $\mathcal{A} = \{A_1, \ldots, A_n\}$, with associated domains D_i. The set of all possible databases over $\mathcal{A}$ is denoted by $inst(DB)$.

3.2.1   Basic Mining Algorithms

Simple mining algorithms can be described with just a few basic aspects:

Description Language: each mining algorithm needs a way to describe the models it is considering. This set of models is called the *description language* in KESO[2]. For classification trees, the description language consists

---

[2]Model language would perhaps have been a better name, but description language was the name used in [44].

of all possible trees.

Quality Measure: given a collection of models, we still have to define what models are good models this is done using a *quality function*. Often one can think of two different types of classification functions. One that assess the quality of a model on new, unseen, tuples such as the *classification accuracy* of a classification tree. The other a *local* measure that us used in the actual construction of the model, such as the entropy measure for ID3 above. It is this latter class of quality functions that is important for the present discussion.

Search Algorithm: given the collection of possible models and a quality function, all the mining algorithm has to do is to find the high(est) quality models. For this, search algorithms are employed, these consist of two parts:

 Search Operators: during the search, new models are created from already explored models, this is done using *search operators*. In the case of association rules, the search operator creates only those candidate frequent sets all of whose subsets are already frequent. In the case of classification trees, the search operator creates trees with one extra split point.

 Search Strategy: whereas the search operator is strongly connected to the particular description language, the search strategy is far more abstract. It simply embodies the strategy through which the search space is traversed. It can vary from exhaustive search through simple hill-climbing to advanced techniques as simulated annealing and evolutionary strategies.

Description Languages It is not very well possible to give an abstract definition of all possible description languages. Instead I'll give some examples.

A *simple description* ϕ for DB is a selection query[3] on DB. A *simple description language* Φ for DB is a set of simple selections. More down to earth, one could see simple descriptions as expressions of the form:

$$A_1 = v_1 \wedge \cdots \wedge A_j = v_j; \{A_1, \ldots, A_j\} \subseteq \mathcal{A} \wedge v_j \in D_j$$

These simple descriptions (also known as *attribute value* selections) form the basis of many other description languages. Simple examples are rules, which have a simple description both at the left and the right hand side.

A slightly more complex example is given by a set description language. A subset Ψ of $\mathcal{P}\Phi$ is a *set description language* for DB. An element $\psi \in \Psi$, i.e., a set of elements of Φ is a *set description*.

Such set description languages can, e.g., be used to define classification trees.

---

[3]For a precise definition of a *query*, consult, e.g., [1]. A selection query is a query whose result is a subset of the original table.

Quality Functions Quality functions are even harder to define abstractly, but their communality is that they are ultimately based on counts on the database. We will return to this topic later in this paper while discussing *datacubes*. For simple descriptions as defined above, a quality function Q determines the quality of a description $\phi \in \Phi$ based on a *two-way table* over the *support* of ϕ. The support of ϕ is simply that part of the database that is described by ϕ. In other words, the support of ϕ, denoted by $\langle \phi \rangle$ is defined by (note we use square brackets to denote multi-sets):

$$\langle \phi \rangle_{db} = [t \in db \mid \phi(t)]$$

Two way tables have sets of source and target attributes, denoted by S and T with $S, T \subseteq \mathcal{A}$. The two-way table has as schema $S \cup T \cup \{Count\}$, where $Count$ has the natural numbers as domain. The tuples in the two-way table over the support of ϕ are those tuples τ that satisfy:

1. $\exists t \in \langle \phi \rangle : (\forall A \in S \cup T : \pi_A(t) = \pi_A(\tau))$

2. $\pi_{Count}(\tau) = |[t \in \langle \phi \rangle \mid \forall A \in S \cup T : \pi_A(t) = \pi_A \tau]|$

The target attributes T are parameters that are set by the quality function; in other words, T is constant during a data mining task. The source attributes S denote a context, S is variable during the execution of the task, its value is set by the search strategy. For given parameters S, T, ϕ, and db, the two-way table is denoted by $2WT(S, T, \phi, db)$

The quality function Q further computes some aggregate function F over the two-way table, which yields the final quality.

For example, $2WT(age, damage, true, db)$ evaluates (conceptually) to the familiar kind of table (on some hypothetical database):

	damage = yes	damage = no
age = 16	54	16
age = 17	54	36

Search Algorithms As explained above, the search algorithms consist of two components, viz., search operators and search strategies. Sometimes, as with association rules, the strategy is exhaustive search, most often it is a heuristic strategy. An example of such a heuristic strategy is the hill-climber, which can be schematically specified by

Hill Climber
    Current:= initiate
    Neighbours := neighbour(Current)
    While $\exists n \in$ Neighbours : quality(n) > quality(Current) **do**
        Current := max-quality(Neighbours)
        Neighbours := neighbour(Current)

Initiate starts the search in a (perhaps random) place. The *neighbour* operator is the search operator that computes all *neighbours* of the current point and *quality* is the quality function. In other words, the hill-climbing strategy finds a local maximum by starting at a random point, searching in the immideate neighbourhood for a (maximally) higher point until such points no longer exist.

The immideate neighbourhood is defined with the *neighbourhood operator* on the *description language*. So, for the classification trees this are all trees that have one more split-point.

The hill climbing strategy is a rather naive (it may get stuck in a local maximum) and far more sophisticated search strategies exist, such as *simulated annealing* and *evolutionary strategies*. However, in practice hill climbing and a simple variant the *beam search* are often used. One reason for this is that both are very cheap strategies and the second is that they give rather good results in general.

The *beam search* is a variant of hill climbing in which one doesn't continue from one point in the search space, but from k points. From the single initial point, all neighbours are computed. If at least one of the neighbours is of higher quality, the k best points are made current. That is, the neighbours of all these k points are computed. ¿From then on, we continue with the best k neighbours untill there are no more points that are better than our current points.

To sum all of this up, note that by defining operators on the description lnguage language we turn it into an *algebra*. That is, a description algebra is specified by a pair $(\Phi, \mathcal{O})$, in which Φ is a description language and $\mathcal{O}$ a set of operators on Φ. The definition of a basic mining algorithm is now given by:

> A basic mining algorithm is a triple $((\Phi, \mathcal{O}), Q, S)$, such that $(\Phi, \mathcal{O})$ is a description algebra, Q is quality function, and S is a search strategy that uses only operators defined in $\mathcal{O}$.

3.2.2   Mining Algorithms

Basic mining algorithms are only part of the story. Most mining algorithms are far more elaborate, e,g., to avoid overfitting or to handle continuous attributes. For an example of the former, consider the classification trees of Section 3.1. It was already noted there that the resulting tree has to be pruned to avoid overfitting. If the tree is grown until it fits the data set exactly, the chances are that it will perform badly on new cases. By pruning away insignificant branches, the risks of overfitting are minimised. The ID3 algorithms as described in Section 3.1 can only handle discrete attributes. Clearly, splitting a continuous attribute on all values it takes is ludicrous. Rather, while considering a continuous attribute we should look for a good discretization.

In both cases we start a new search. In the pruning case for branches that can be pruned, in the discretization case for good discretizations. In other words, one builds complex mining algorithms from basic mining algorithms. For example, a variant of ID3 including continuous attributes could be programmed schematically as follows:

ID3 Variant
   Perform ID3 algorithm

If considered attribute A is continous
  Discretize(A)

In which Discretize is a separate mining algorithm that looks for good discretizations. All mining algorithms in KESO are built in this way from basic mining algorithms.

4   KESO: a Data Mining System

4.1   The Data Mining Kernel

The components of a (basic) mining algorithm are faithfully reflected in the architecture of the KESO system. In particular, S is mirrored by the *Search Manager*, $\mathcal{O}$ by the *Description Generator*, and Q by the *Quality Computer*. Obviously, there is not one component that embodies the description language Φ, since this is the "hypothesis language" shared by the components.

4.1.1   The Search Manager

The Search Manager contains a number of·Search Modules, each of which implements a search algorithm[4]. That is, each of these modules can orchestrate a search process, but it does not manipulate descriptions nor does it compute qualities. It relies on other components to perform these tasks. It uses the results of these tasks to decide what the next step in the discovery process should be.

4.1.2   The Description Generator

The Description Generator also consists of a number of modules. Each of these modules implements one of the operators on a specific descrition language. These modules may consult the knowledge base to determine what the result of a operation on a (set of) descriptions should be. A search component will tell the description generator what operators it should apply to which (sets of) descriptions.

4.1.3   The Quality Computer

The Quality Computer is the component that interfaces with KESO's mining server. Through this server it queries the database to gather statistics on a description, such that it can compute the quality of this description. It contains, modularly, a set of different quality functions. All of these sub-modules share the same interface to the dataserver.

---

[4] Actually, the KESO architecture does not contain a Search Manager, the Search Modules simply share a library, e.g., for communication purposes. Conceptually, it might be easier to think in terms of a Search Manager.

The quality computer computes the quality of the descriptions that are newly generated by the description generator. The set of new descriptions together with their quality are handed back to the search module that bases its next step on this information.

4.1.4 The Communication

The communication of these components is through the Search Space Manager, which persistently stores the search space. The logical flow of information is as sketched above, albeit through the Search Space Manager. That is, the components store records in the Search Space Manager and request records from it. The records in the Search Space Manager, contain, a.o.,

- an *id-field*, i.e., logical name for the description,

- a field for the description itself,

- a field for the quality of this description,

- a field which contains the name of the operator using which the description is generated,

- a *parent* field which contains the id(s) of the descriptions from which the description is generated.

The parent field gives the search space a graph-structure. Each search space has a default element, called *anchor*, on which the search graph is anchoredred. If requested, the quality of anchor record is smaller than any other quality, e.g., it is $-\infty$, and it is its own parent.

One of the motivations for the persistent storage of the Search Space is that it enables the user interaction. Others are that it allows the user to roll-back in a discovery session and to return to such a session at a later date.

4.1.5 The Mining Conductor

As described in the previous Section, mining algorithms are built from basic mining algorithms. The Search Manager, The Description Generator and the Quality Computer implement the basic mining algorithms. The *Mining Conductor* is the final piece of the puzzle: it activates the correct modules and sets up the proper communication channels (i.e., it instantiates a search space database for the task at hand).

More importantly, the mining conductor allows for the implementation of (complex) mining algorithms. For, as stated before, these complex mining algorithms are built up from the basic ones. It is perhaps easiest to see each mining algorithms as a finite state automaton in which each node implements a basic mining algorithm. With this view, the mining conductor simply executes this finite state automaton. In reality the situation is slightly more complex, but a detailled discussion of that is outside the scope of this paper.

In ending this subsection on implementation issues, note that each sub-process in a mining algorithm gives rise to its own search space. These search spaces are maintained seperately, although the relationships are also maintained. The main reason is that this ensures relatively simple databases that are easily searcheable by the system and the user alike.

4.2   Supporting the KDD process

The KESO project was limited both in time and money. For this reason, the interface of the KESO system gives good support for mining but the support for the complete KDD process is rather rudimentary. Still, the architecture was set up in such a way that such support could easily be integrated. In fact, that is what Data Distilleries, the commercial partner in the project has done. In this subsection we give a brief overview of this intercae, called Expert/Surveyor to give the reader an idea how a system can support the KDD process. For more details the reader is referred to the website at datadistilleries.com.

Data Mining in Expert/Surveyor is organized along projects. One project may have many mining questions, but they all have to use the same mining table. The construction of this table is supported by Expert/Surveyor.

A project is visualised in the interface by a project window. Such a project window consists of different layers. Bottom up, these are the following:

Source Backend   In this layer the user clarifies the types of the data sources that are to be used in the mining. The supported types at the moment are ORACLE databases, file systems and ODBC compliant DBMSs.

Source Data Sources   In this layer the user gives the actual sources to be used. So, if ORACLE is used, the correct ORACLE databases are selected.

Source Data Sets   In this layer, the user selects which data sets within the sources are actually used. That is, which tables in the database or which files in the file system are to be used.

Source Schema   In this layer, the schema that connects the different source data sets is specified. That is, it gives the logical tables (i.e., the source data sets) and their relationships as customary in many database design formalisms. Note, that as far as these details are known within the sources, this schema is generated automatically. For data sets from different sources, one has to add such links by hand.

While working in this layer, the user can at any time view the properties and the contents of that data set. So, one can see which attributes a table has, the tuples in that table, but also the distribution of the values of an attribute, e.g., via histograms.

A very important feature of this layer is that one can add new, derived, attributes to existing tables. The expression editor is the tool with which these new attributes can be created.

As a final step in this layer one has to create a mining table since the algorithms in Data Surveyor currently all expect only one table. To do this, one cab create a new table in which all the relevant attributes are copied (note that the implicit joins are computed along the links selected by the user).

Mining Questions In this layer, the user can formulate all his mining questions. This is done by entering the algorithm, the quality, the mining table, and the (optional) selection on that mining table. Note that such a mining question could consist of only one step in a complete algorithm. That is, the user can mine interactively.

Mining Answers By executing mininq questions, mining answers are generated. Interactive mining means that one switches back between this layer and the previous one. One determines a question, inspects the results and then extends the initial question, and so on.

Models All the intermediate results that are found can be annotated. In fact, all the steps taken so far have plenty of scope for annotation. By annotating steps on the way, the user automatically creates the documentation of his mining process.

The results that the user wants to be used can be selected and copied to the model layer. For specific applications, there are layers that allow the user to export these models to that application.

5 Algorithms in KESO and Beyond

Now that we have introduced mining algorithms and the framework how these are implemented in the KESO system, it is time to discuss some of the algorithms and their implementation in KESO. Moreover, we'll briefly discuss how two well-known statistical algorithms could be implemented within the system. The purpose of this latter discussion is mainly to illustrate the generality of the ideas underlying KESO.

5.1 Association Rules and Classification Trees

The first algorithms to describe in KESO are, of course, association rules and classification trees. In fact, since these algorithms have described in some detail already, their discussion here will be brief.

5.1.1 Association Rules

As explained in Section 3, association rules are discoverd in two steps. First the frequent sets are discovered, and then the association rules are generated from these frequent sets. The implementation in KESO is a straight forward application of this idea.

The search strategy for frequent sets is simply exhaustive search, that is, in principle all subsets of articles are generated. The pruning of the search space is done by the search operator: only those subsets that could be frequent (as explained in Section 3) are generated.

The generation of association rules from the frequent sets is again an exhaustive search. Now the search operator prunes nothing, but simply generates all possible alternatives and the validity of the rule is tested.

5.1.2   Classification trees

Here we'll give some more details to illustrate (yet again) the philosophy underlying the KESO system. We restrict ourselves to ID3 rather than C4.5 which is actually implemented in KESO. This restriction simplifies our discussion.

First of all, we have to define the description language. From the observation above, we see that each description has to represent a complete tree. That is, we have to use set descriptions. In fact, we use the set-descriptions given as an example in Section 3.2.1. Each simple description ϕ that is an element of such a set description ψ simply denotes the slection along the path from the root of ψ to a leaves. Clearly, not all such set descriptions can be interpreted as a proper tree. Hence, we have to ensure that we only construct set descriptions that do represent a proper tree.

Next, we have to specify the search strategy that is used. ID3, and in fact almost all tree algorithms use one of the simplest of all: the hill climber we already discussed in Section 3.2.1

The Hill-Climber algorithm as specified there is not especially tailored towards the construction of decision trees. That is achieved in the specification of the operators.

The first operator is the *initiate* operator. This operator should construct a one-node tree, in which the single node represents the complete database. The selection along the empty-path is simply *true*, thus the initial tree is described by the set containing the empty path.

The next operator is the *neighbour* operator. It receives a tree as input and outputs all trees that have one *internal* node extra. That is, we make one split extra. In the case of ID3, the input tree is a set of simple descriptions $\Phi = \{\phi_1, \ldots, \phi_k\}$. And making a split in Φ means that we replace one of the ϕ_i by all selections of the form $\phi_i \wedge A_j = v_j$ such that A_j is not yet "used" in ϕ_i and $v_j \in D_j$.

To make this slightly more precise, denote by Att_ϕ the attributes that are used in ϕ. For example, if $\phi \equiv A_1 = v_1 \wedge A_2 = v_2$, then $Att_\phi = \{A_1, A_2\}$. But paths should be extended with selections on attributes *not yet* used: define $Att^\phi = \mathcal{A} \setminus Att_\phi$. Continuing our example, $Att^\phi = \{A_1, \ldots A_n\} \setminus \{A_1, A_2\} = \{A_3, A_4, \ldots, A_n\}$. Finally, for a simple description ϕ and an attribute $A \in Att^\phi$, the *split* in ϕ on A is given by:

$$split(\phi, A) = \{\phi \wedge A = v | v \in Domain(A)\}$$

With these definitions, the neighbour operator for ID3 is specified as follows:

$$Neighbour(\Phi) = \{\Psi | \Psi \setminus \Phi = \{\psi\} \wedge \Phi \setminus \Psi = \{\phi\} \wedge \exists A \in Att^{\phi} : \psi \in split(\phi, A)\}$$

For the quality function, note that a node N in ID3 corresponds with a description ϕ that belongs to the path that ends in N. The set S covered by N is thus $\langle \phi \rangle$. Clearly, the class-entropy $Ent(\phi)$ can be computed from a 2-way table . Using the parent in the search space, $E(N, A)$ can also be computed using 2-way tables.

5.2 Interesting Subgroups

Another group of algorithms in KESO is targeted at the discovery of interesting subgroups. A detailed discussion of this group of algorithms can be found elsewhere in these proceedings in the paper by Willi Kloesgen. Here we'll restrict to the basic aspects of these algorithms.

Interesting subgroups are parts of the database that "behave" differently from the rest of the database. This behaviour is assumed to be encoded in a *target attribute* (a dependent variable in Statistics) similar to the class attribute for classification problems. For this discussion we assume that the target attribute is binary, i.e., that it only takes the values 0 and 1. Such an attribute could, e.g., encode whether customers own a certain product, responded to a mailing, or claimed on an insurance policy in a given year. The goal is then to find subgroups with a high (or low) proportion of ownership (i.e., of 1's).

The description language used in this case is that of the simple descriptions of Section 3.2.1 with the proviso that for continuous attributes the selection is always on an interval rather than a single value. That is, descriptions of the form:

$$A_{i_1} = v_{i_1} \wedge \cdots A_{i_k} = v_{i_k} \text{ with } \begin{cases} v_{i_j} = c & \text{if } \mathrm{Dom}(A_{i_j}) \text{ is discrete,} \\ v_{i_j} = [c_1, c_2] & \text{if } \mathrm{Dom}(A_{i_j}) \text{ is continuous.} \end{cases}$$

Such descriptions denote a subset of the database by simply interpreting them as a selection condition.

In principle, the quality of such an expression is simply the fraction of 1's its subgroup has for the target attribute. However, we assume that the database is only a sample of the population. Therefore, not the fraction is taken as a quality measure, but a *confidence interval* around this fraction with a user defined confidence level C . That is, an interval $[c_1, c_2]$ such that given the size N of the subgroup and the fraction f of 1's of that group the probability that the fraction p of 1's is outside the interval is smaller than $1 - C$, i.e.,

$$Prob(p \notin [c_1, c_2] | f, N) < 1 - C.$$

Note that there may be many intervals with this property. Whenever possible, we take the interval that is symmetric around f. In the other cases, we take the interval that starts at 0 or ends at 1.

To compute these intervals, note that we can see the values of the target attribute as the outcome of a Bernoulli experiment. That is, the observed number of 1's follows a Binomial distribution. In other words, we can compute the interval using the inverse cumulative density function for of the Binomial distribution as implemented, e.g., in cdflib (see, e.g., www.stat.cmu.edu) .

The search strategy is a variant of the hill-climber called *beam search*. The main difference is that the beam search explores a user specified number of paths (the *width* of the search) at the same time. It is specified as follows:

Beam Search(width)
  Current:= { initiate }
  Neighbours := neighbours(Current)
  While $\exists n \in$ Neighbours : quality(n) > quality(Current) **do**
    Current := best(width)(Neighbours)
    Neighbours := neighbours(Current)

In other words, like the hill climber the beam search starts in one point. In all subsequent steps, the best *width* number of neighbours is used for further exploration.

One question is: when is a new description ψ better than its parent ϕ? There are two obvious solutions and both are supported by KESO. The first is that the confidence interval of ψ is disjunct and higher than that of ϕ; the larger the distance between these two confidence intervals the better ψ is. The other solution doesn't compare with ϕ, but with the complement of ψ in ϕ, i.e., with the confidence interval for $\phi \setminus \psi$.

The second question is: how are the neighbours computed? The first step in the definition of this operator is easy; it is the mapping of a *neighbour* operator over the set of current descriptions. That is,

$$\text{neighbours(Current)} = \{\text{neighbour}(c)|c \in \text{Current}\}.$$

If all the attributes are discrete, this *neighbour* operator is very similar to that for trees: a new attribute value selection is added to the description. That is,

$$\text{neighbour}(\phi) = \{\phi \wedge A_i = c_i | A_i \in Att^\phi \wedge c_i \in \text{Dom}(A_i)\}.$$

For, continuous attributes, however, we do not want a single value but an interval. To get these intervals, $\text{Dom}(A_i)$ has to be discretized first. One way of doing that is to discretize all continuous attributes before the search starts and use these discretized attributes rather than their continuous counterparts. However, it is very well possible that a discretization of, say, *Age* should be different for *women* than for *men*.

To facilitate such differences, discretizations are done on the fly. For the simplicity of the specification, assume that our discretization operator simply returns the domain of an attribute if that attribute is discrete. The neighbour operator is then specified by:

$$\text{neighbour}(\phi) = \{\phi \wedge A_i = c_i | A_i \in Att^\phi \wedge c_i \in \text{discr}(\phi)\text{Dom}(A_i)\}.$$

Note that the *discr* operator gets the current description ϕ as an extra parameter to ensure the on the fly discretization.

So, how are continuous domains discretized? Obviously, discretization is a seperate mining algorithm. The description language is fairly simple: it simply consists of all possible discretizations of a given domain. That is, the set of all sets X of subsets of $\text{Dom}(A_i)$ such that

1. $\bigcup_{x \in X} x = \text{Dom}(A_i)$

2. $i \neq j \rightarrow x_i \cap x_j = \emptyset$

The quality of a discretization is computed as the weighted sum of the distances between the confidence intervals (for the fraction of 1's of the target attribute) of adjacent intervals. The weight is simply the number of intervals in the discretization. The search for a good discretization is performed by iteratively binary splittings starting with the whole domain; if the confidence interval for two adjacent intervals overlap, they are united.

5.3   Bayesian Networks

Bayesian networks are way to see the dependencies between the attributes in the database. To do this, the database is seen as a joint probability distribution over the attributes $p(A_1, \ldots, A_n)$. This distribution gives the probability (relative frequency) of the occurrence of a certain tuple $(v_1, \ldots, v_n)$ in the database by $p(A_1 = v_1, \ldots, A_n = v_n)$.

In database theory *normalization* of the database is a well-known process. A relation R is split into two (or more) relations, say R_1 and R_2, such that $R = R_1 \bowtie R_2$. It simplifies updates by removing redundancy in the database. Something similar can be done with joint probability distributions using *Bayes Theorem*, e.g.,

$$p(A_1, \ldots, A_i, A_{i+1}, \ldots, A_n) = p(A_1, \ldots, A_i | A_{i+1}, \ldots, A_n) p(A_{i+1}, \ldots, A_n)$$

If there are multiple values for $A_1, \ldots, A_n$ that occur with the same value for $A_{i+1}, \ldots, A_n$ this can be used as a less redundant representation, just as normalization. Far more insight, however, can be gained if $A_1, \ldots, A_i$ are *independent* of $A_{i+1}, \ldots, A_{n-1}$ given A_n for then we have:

$$p(A_1, \ldots, A_i, A_{i+1}, \ldots, A_n) = p(A_1, \ldots, A_i | A_{i+1}, \ldots, A_n) p(A_{i+1}, \ldots, A_n)$$
$$= p(A_1, \ldots, A_i | A_n) p(A_{i+1}, \ldots, A_n)$$

In fact, by repeatedly using Bayes Theorem and exploiting conditional independencies we could, e.g., end up with the following equation:

$$p(A_1, \ldots, A_n) = p(A_1 | A_2) p(A_2 | A_3) \cdots p(A_{n-1} | A_n).$$

Clearly, the righthand-side of this equation provides far more insight in the data than the lefthand-side.

It should be obvious that we can always use Bayes Theorem to arrive at an equation of the form:

$$p(A_1, \ldots, A_n) = p(A_1|\pi_1)p(A_2|\pi_2) \cdots p(A_n|\pi_n)$$
$$= \prod_{i=1}^{n} p(A_i|\pi_i)$$

in which $\pi_i \subseteq \{A_1, \ldots, A_{i-1}, A_{i+1}, \ldots, A_n\}$; note that some (or all) of the π_i may very well be empty. Moreover, without loss of generality, we may assume that π_i contains no superfluous attributes. That is, we may assume that there are no (sets of) attributes X in π_i such that A_i is conditionally independent of X given $\pi_i \setminus X$. In such an equation, π_i is called the *parent set* of A_i.

As an aside, note that there may be different ways to factorize the original joint probability distribution in product of conditional (and marginal when $\pi_i = \emptyset$) probability distributions. The simpler the factorization is, the more insight it provides.

Such a factorization can be visualized as a directed graph, by making the attributes the nodes of the graph and drawing an arrow from each *parent* in π_i to A_i. Since we use Bayes Theorem to construct the factorization and since conditional independencies will at most remove elements of the parent set, a little reflection shows that such a representation does not contain cycles. In other words, it is a DAG (= Directed Acyclic Graph). If we annotate each node A_i with its conditional probability distribution $p(A_i|\pi_i)$ we have a *Bayesian Network*.

More formally, a Bayesian Network a pair (B_S, B_P) in which B_S is a DAG and B_P is a set of conditional probability distributions, such that for each node $N \in B_S$ there is a distribution $P(N, \pi_N) \in B_P$, where π_N coincides with the parents of N in B_S.

Bayesian networks are not only useful because they provide insight in the conditional (in)dependence structure of the database, they also allow one to reason with possibly incomplete information. A discussion of this aspect is beyond the scope of this paper, the reader is referred to [27, 39] for more information of this topic.

Now that we know what Bayesian networks are, it is time to discuss how they are discovered from the database. It should be clear that we only have to worry about the underlying DAG, from that and the database the accompanying probability distributions are easily estimated. So, how is this graph discovered?

Unfortunately, as for Classification Trees, there is no easy way to compute the best possible network. Unlike that case, more is known about the complexity. For example, it is known that even if the each node has no more parents than k, finding the best network is NP-complete if $k > 1$; see [9]. So, we will have to use heuristic techniques to find a (hopefully) reasonable approximation of this best solution.

In such a heuristic scheme, it is attractive to use a hill-climber or a simple variant of that such as a beam search or simulated annealing as a search strategy.

For example, we would start with the empty network (i.e., just nodes and no (directed) edges at all) and at each step of the search, we would consider to:

1. add an arc, or

2. reverse an arc, or

3. remove an arc.

The advantage of such an approach is that we consider at most a polynomial number[5] of networks out of the exponential number of possibilities.

To do this, we need a quality measure that allows us to compare how well two marginally different Bayesian networks representations represent the original database. A straight forward idea may seem to lie in the probability distributions. For, both from the database and from the Bayesian network we can estimate a (joint) probability distribution. Moreover, there are well-known measures to estimate how close two probability distributions are to each other, e.g., the Kullback-Leibler measure (an entropy measure). Let db_u be the database from which all duplicate tuples have been removed, then this measure is defined for two such probability distributions p_1 and p_2 by:

$$\sum_{t \in db_u} p_1(t) \frac{p_1(t)}{p_2(t)}$$

The closer this measure is to 0 the more p_1 and p_2 resemble each other.

There are, however, two disadvantages with this approach. Firstly, the Kullback-Leibler measure favours the fully connected network. For, in that case, since no redundancies have been removed, the estimated probability from the network will be the same as that estimated from the database. This hardly provides insight. The second disadvantage is that the Kullback-Leibler measure is that it is non-local. That is, even if network n_2 is just a small variation of network n_1, knowing the estimated probability distribution $\hat{p}_1$ for n_1 doesn't help in estimating $\hat{p}_2$ for n_2. Estimating a full joint distribution is a costly problem. A local measure would be far more efficient.

If all the attributes are discrete and the database contains no missing values, such local measures exist under some mild assumptions; see [10] for full details, we'll only discuss some of the major aspects.

The Bayesian approach is that we want to find the network structure that is most likely given the database. Bayes rule tells us:

$$P(B_S, db) = \frac{P(G, db)}{P(db)}$$

So, for two network structures B_{S_1} and B_{S_2} w have:

$$\frac{P(B_{S_1}|db)}{P(B_{S_2}|db)} = \frac{\frac{P(B_{S_1}, db)}{P(db)}}{\frac{P(B_{S_2}, db)}{P(db)}} = \frac{P(B_{S_1}, db)}{P(B_{S_2}, db)}$$

---

[5]polynomial in the number of attributes

In other words, we "only" have to know $P(B_S, db)$ for a quality measure of the network. We can do this by integrating over all possible value assignments of B_P:

$$P(B_S, db) = \int_{B_P} f(B_S, db, B_P) dB_P$$

Applying the Bayes Theorem recursively, we get:

$$P(B_S, db) = \int_{B_P} f(db|B_S, B_P) f(B_P|B_S) P(B_S) dB_P$$

Under the assumption alluded to above, further evaluation of this expression yields:

$$P(B_S, db) = P(B_S) \prod_{i=1}^{n} \prod_{j=1}^{q_i} \frac{(r_i - 1)!}{(N_{ij} + r_i - 1)!} \prod_{k=1}^{r_i} N_{ijk}!$$

with, $P(B_S)$ the prior probability of B_S, n the number of attributes, q_i the cardinality of the domain of the parent set π_i (i.e., the cardinality of the cartesian product of the domains of the elements of π_i), r_i the cardinality of the domain of A_i, N_{ijk} the number of cases in the database for a given value $v_i = x_{ik}$ of attribute A_i and given values $x_{\pi_i j}$ for the parent attributes, and $N_{ij} = \sum_{k=1}^{r_i} N_{ijk}$.

For computational purposes, the Bayesian quality, $B(B_S, db)$ of a network B_S is defined as $log(P(B_S, db))$. Under the assumption of an uniform prior distribution, this can be computed as:

$$
\begin{aligned}
B(B_S, db) &= log(P(B_S) \prod_{i=1}^{n} \prod_{j=1}^{q_i} \frac{(r_i - 1)!}{(N_{ij} + r_i - 1)!} \prod_{k=1}^{r_i} N_{ijk}!) \\
&= log(P(B_S)) + log(\prod_{i=1}^{n} \prod_{j=1}^{q_i} \frac{(r_i - 1)!}{(N_{ij} + r_i - 1)!} \prod_{k=1}^{r_i} N_{ijk}!) \\
&= log(P(B_S)) + \sum_{i=1}^{n} \sum_{j=1}^{q_i} log(\frac{(r_i - 1)!}{(N_{ij} + r_i - 1)!}) + \sum_{k=1}^{r_i} log(N_{ijk}!)
\end{aligned}
$$

In other words, if we define the Bayesian quality of a node A_i in B_S for db by:

$$m^B(A_i, B_S, db) = \sum_{j=1}^{q_i} log(\frac{(r_i - 1)!}{(N_{ij} + r_i - 1)!}) + \sum_{k=1}^{r_i} log(N_{ijk}!)$$

Then:

$$B(B_S, db) = log(P(B_S)) + \sum_{i=1}^{n} m^B(A_i, B_S, db)$$

Let B_{S_1} and B_{S_2} be two networks that differ only in the parent set of A_1, e.g., $B_{S_2} = B_{S_1} \cup \{A_2 \to A_1\}$ Then:

$$B(B_{S_2}, db) - B(B_{S_1}, db) = m^B(A_1, B_{S_2}, db) - m^B(A_1, B_{S_1}, db)$$

In other words, in the search process we only have to compute the local measure $m^B(A_1, G_1, db)$ as promised above. With this measure, our heuristic search algorithm for Bayesian networks is defined.

5.4   Statistical Algorithms

¿From the discussion of algorithms in KESO the reader may get the impression that the framework works well for Machine Learning type of algorithms but does not support more statistical algorithms. In one sense that is true, the KESO system has not been built to support the parameter estimating procedures common to many statistical algorithms. In another sense, however, it is not true. Exploratory data analysis algorithms from Statistics fit well into the framework. To illustrate this, we briefly discuss two such algorithms, viz., stepwise regression and projection pursuit.

5.4.1   Regression

One of the best-known techniques in Statistics is without doubt regression analysis. In formal terminology, let (X, Y) be a pair of random variables such that X is $\mathcal{R}^d$ valued while Y is $\mathcal{R}$ valued. The problem is to estimate the *response surface*

$$f(x) = E(Y|X = x)$$

from n observations $(X_1, Y_1), \cdots, (X_n, Y_n)$ of (X, Y).

A simple way to fit a function to these n observations is through *least squares estimation*. First a parametric form for f is chosen, e.g., if f is assumed to be a linear surface, we have $f(\vec{x}) = \sum_{i=1}^{n} a_i x_i + a_o$. Following, the parameters a_i are estimated by minimising

$$\sum_{i=1}^{n} (Y_i - f(X_i))^2.$$

Defined in this way, it seems far from our data mining algorithms, there is no search at all. However, what happens if our goal is simply to predict Y and the X simply contains all other measurements about the objects that are known at prediction time. That is, what happens if not all attributes in X are necessary for the prediction of Y.

Classically, the analyst ought to define the proper functional expression for f, which includes using the correct attributes. However, we could let an algorithm decide which attributes the include?

The quality of a regression function f is often characterized by the *coefficient of determinination* R^2 which gives the proportion of the variation in Y that is explained by f. It can, e.g., be computed as the square of the correlation between Y and $f(X)$. This may seem an ideal quality measure for a simple heuristic search:

Start with the empty expression:
Repeat consider all expressions with one attribute more
    Choose the one with the highest R^2 value
    Until R^2 doesn't rise anymore

There are, however, two problems with this simple approach. Firstly, R^2 can only rise if we add extra attributes. This problem can be alleviated using the *adjusted* R^2, which is defined as:

$$1 - \frac{n-1}{n-p}(1 - R^2)$$

for a regression equation with p attributes computed over a database with n tuples.

The second problem is, it is nice if the (adjusted) R^2 rises, but is the contribution of all attributes (still) significant. Clearly, it doesn't make sense to add non-significant attributes. To test whether R^2 is significantly bigger than 0, one can use an F statistic:

$$\frac{\frac{\sum(y_i - \hat{y}_i)^2}{p-1}}{\frac{\sum(y_i - \bar{y})^2}{n-p}}$$

with n and p as above. This statistic has (p - 1, n - p) degrees of freedom. So, to test whether an attribute is (still) significant, we compute a partial F value by first removing the effect of the other variables.

With these two adjustments, our simple algorithm above is known as stepwise regression. Obviously, it fits the KEso framework. There are, however, two cautions one should be aware of, when applying this technique in a data mining context. Firstly, if n is large (as is often the case in a data mining context) the adjusted R^2 will quickly be the same as R^2 itself. Secondly, if n is large the F-statistic will almost always seem significant. This can be alleviated by using a (small) sample from the database, or using a very high significance level, and, of course, keeping the expert in the loop.

5.4.2   Projection Pursuit

Mapping multivariate data into low dimensions for visual inspection is a commonly used technique in data analysis; if only because of the uncanning ability of humans to discover structure in two-dimensional plots. The discovery of such mappings that reveal the salient features of the multidimensional data set is in general far from trivial. *Projection Pursuit* (PP) introduced by Friedman and Tukey in [18] is a technique to discover such mappings.

In a nutshell, PP works as follows. We have a p-dimensional dataset X and we examine "all", say, two-dimensional projections of X. We are given some quality function, called the projection index, with which we calculate the quality of all the projections. PP then reports the projection with the highest quality.

The simplest mappings from higher to lower dimensions are, linear, projections. That is, linear maps A of, say, rank 1 or 2. By definition, PP searches for a projection A that maximises a quality function, in this context it is called the *projection index*.

To get more concrete, let X be a $\mathcal{R}^d$ valued random variable and let $\mathcal{X} = \{X_1, \cdots, X_n\}$ be a set of n observations of X. A 1-dimensional projection A is then a $1 \times d$ matrix of rank 1. The quality of A should be determined from the data set $A(\mathcal{X}) = \{AX_1, \cdots, AX_n\}$.

Many projection indices are possible, an important observation by Huber [26] is that the index should measure how far the projection is away from a set of data points sampled under a normal distribution. The heuristic arguments underlying this claim are:

- A multivariate distribution is normal iff all its one-dimensional projections are normal. Thus, if the least normal one dimensional projection is normal, we need not look at any other projection.

- For most high-dimensional data sets most low-dimensional projections are approximately normal.

A simple projection index in this case is, thus, a χ^2-test. Another example is the sample entropy, i.e.,

$$\frac{1}{n} \sum_{i=1}^{n} \log(\hat{f}(AX_i))$$

in which $\hat{f}$ is the density estimate of the projected points. Friedman and Tukey's original index I is the product of two functions s and k, where s measures the spread of the data and k describes the "local density" of the data after projection.

Defining the index is only part of the work. The question is, of course, how to find the projection A that maximizes the index. Friedman and Tukey mention that their projection index is sufficiently continuous to allow the use of hill-climbing algorithms for the maximization.

¿From this, brief, description of Projection Pursuit, it should be clear how it can be implemented in KESO. The description language consists of all projection planes, the quality function is the PP-index and as a search strategy one can take the hill-climber.

6   Data Mining and Databases

An important observation in the runtime behaviour of KESO (and most probably true in all data mining systems) is that by far the most time is spend in the computation of qualities. That is, the bulk of the work is in the database. So, if one wants to speed-up data mining, one should speed-up the database access.

6.1 Database Support for Data Mining

The interaction between the KESO system and the DBMS is restricted to the computation of the two-way tables. These tables are easily expressed in SQL by:

SELECT Source, Target,
    COUNT(Source, Target) **AS** Count
FROM *db*
WHERE ϕ
GROUP BY Source, Target

So, one way to speed-up the data mining process would be to pre-compute these tables, that is to compute a *data cube* as defined in [19]. However, there are two major disadvantages with this approach. The first is caused by the "on the fly" discretization in KESO. Each such discretization would add another dimension to the cube, thus requiring the computation of a new cube, of which the old cube is only a sub-cube. Computing this new cube is far more expensive than computing the two-way table.

The second disadvantage is that (if there are no continuous attributes) the cube would give the two-way tables for the complete search space. Above we already mentioned that this search space is often far too large to explore completely. In other words, computing the cube is computing far too many two-way tables and would thus take far too much time.

The approach taken in KESO uses a lattice structure on the two way tables, similar to the lattice in [22]. Note that $2WT(S_1, T, \phi_1, db)$ can be computed from $2WT(S_1 \cup S_2, T, \phi_1 \vee \phi_2, db)$ as follows. Project the S_2 attributes out of $2WT(S_1 \cup S_2, T, \phi_1 \vee \phi_2, db)$ and sum the counts of those tuples that become identical. Then select that part of the result that satisfies ϕ_1. This observation induces the following order on two-way tables: $2WT(S_1, T, \phi_1, db) \prec 2WT(S_2, T, \phi_2, db)$ iff $S_1 \subseteq S_2 \wedge \phi_1 \rightarrow \phi_2$. It is obvious that the set of all two-way tables forms a lattice under this order.

This lattice structure is used in two ways in KESO. Firstly, it tells us that if the search algorithm telescopes in on the database (queries smaller and smaller subsets of the database) it is worthwhile to cache intermediate two-way tables since subsequent two-way tables can be computed from these intermediate results. Since the two-way tables are in general far smaller than the database, this speeds-up query processing notably; see [25] for more details and performance figures.

Secondly, KESO sends one batch of two-way queries per "generation" of the search process to the database. These batches contain two-way queries that are closely related. Using the lattice structure, we first compute the two-way table for their smallest common ancestor and derive from that the necessary two-way tables.

The important difference between our usage of the lattice structure and that in [22], is that the authors of [22] use the lattice to compute which views to store once and for all while we take a dynamic approach to view materialisation. We

are dynamic by necessity, since we do not know which queries will be asked. Since most (heuristic) search processes are Markov processes, the best we can do is to stay on the heels of the search process.

A similar observation is that $2WT(S, T, \phi_1 \vee \phi_2, db)$ can be computed from $2WT(S, T, \phi_1, db)$ and $2WT(S, T, \phi_2, db)$ by merging the two tables and sum the counts of identical tuples.

This is important for KESO to run on top of a parallel or a distributed database. It tells us that we can compute the two-way tables on each of the database fragments in parallel and subsequently merge the results; note, this is also observed in [19]. As of yet, we do not have experimental evidence that this speeds up the mining process, but the advantages seem obvious.

6.2   Computing Aggregates in the Database

The motivation for two-way tables as "the" query on which quality functions are based is the fact that all statistics that can be derived from the database can be derived from two-way tables. This follows from the observation that $2WT(\mathcal{A}, \emptyset, true, db)$ simply yields the *set* of tuples in the database extended with their multiplicity.

The disadvantage of using two-way tables is the communication between KESO and the underlying DBMS. Although in practice the two-way tables are far smaller than the database, the overhead is considerable. In other words, KESO would become far more efficient if the evaluation function itself would be computed in the database. In fact, it would be optimal if we could compute the evaluation function while we are constructing the two-way table. What SQL offers in this respect are aggregates beyond **COUNT** and **SUM**.

Allowing such aggregation functions in our two-way tables does offer huge potential savings in the communication between KESO and the underlying DBMS. Generalising the two-way tables to allow the computation of aggregate functions, however, is a potential threat to the optimization schemes outlined in the previous section. In other words, the savings in communication costs could be anihilated by the increase in the costs of computing the tables. Clearly, the aggregated value itself cannot function as the intermediate result that can be re-used for subsequent quality calculations. What should be re-used is the table on which this aggregate is computed. This table itself (the generalisation of the two-way table) may be computed using a different aggregation function, say G. If we want our observations of the previous section to go through, G has to be, [19], *distributive*:

Let $X = \{X_{i,j} \mid i \in \{1, \ldots, I\}, j \in \{1, \ldots, J\}\}$ be a two-dimensional data set. The aggregate function G is distributive if there exists an aggregate function H such that

$$G(X) = H(\{\{G(\{X_{i,j} \mid i \in \{1, \ldots, I\}\}) \mid j \in \{1, \ldots, J\}\})$$

Examples of distributive aggregate functions are **SUM** and **COUNT**.

Clearly, there is no need that F is based on only one distributive aggregate function G, it may depend on a *vector* of such functions provided there is a function M that combines these aggregated values into one aggregate value. In the spirit of [19], we say that an aggregate function F is distributive algebraic if there exists an $k - tuple$ of distributive aggregate functions $(G_1, \ldots, G_k)$ and a function M such that:

$$F(X) = M(G_1(X), \ldots, G_k(X))$$

Examples of distributive algebraic aggregate functions are all the (central) moments of the distribution of attribute values.

Finally, we define a *Data Mining Measure* as an $l - tuple$ of distributive algebraic aggregate functions. Since a distribution is completely characterised by all its central moments, a *Data Mining Measure* is a true generalisation of a two-way table. All the more since the optimisation possibilities for Data Mining Measures are the same as those for two-way tables.

Currently experiments are underway with a version of KESO in which the quality functions are based on Data Mining Measures rather than two-way tables.

7   Conclusions and Acknowledgements

This paper describes the most important concepts and considerations underlying the KESO data mining system. In the three years of the project, the basic architecture has proven its merits. Although the KESO system itself is not available, DATA SURVEYOR from Data Distilleries[6] is a commercially available direct descendent of the KESO system. Moreover, both at CWI and at GMD, the KESO system is in use as a research tool in Data Mining.

If (or better when) I will start with the development of a new data mining tool, many low-level decisions taken in the KESO project will be altered. However, the outline of the architecture as detailed in this paper will largely stay the same.

This paper does not describe all work performed in the KESO project; if only because this would require a book rather than an article. The main developers were Willi Kloesgen, Stefan Wröbel, and Dietrich Wettscherek from GMD, Yka Hutalla, Heikki Mannila, and Inkeri Verkamo from the University of Helsinki, Fred and Donald Kwakkel and Marijn Dee from Data Distilleries, and Robert Castelo, Martin Kersten, and Arno Siebes from CWI. Many related papers can be found via their websites.

The users from CBS (=Statistics Netherlands), FinStat (= Statistics Finland), Forth and Infratest GmbH played an important role in the project; without their criticisms the final system would have been less useful.

---

[6]www.datadistilleries.com

References

[1] Serge Abiteboul, Richard Hull, and Victor Vianu. *Foundations of Databases*. Addison Wesley, 1994.

[2] R. Agrawal, P. Stolorz, and G. Piatetsky-Shapiro, editors. *AAAI-98 Conference on Knowledge Discovery and Data Mining*, New York, New York, 1998.

[3] Rakesh Agrawal, Tomasz Imielinski, and Arun Swami. Mining association rules between sets of items in large databases. In *Proceedings of the 1993 International Conference on Management of Data (SIGMOD 93)*, pages 207 – 216, May 1993.

[4] Rakesh Agrawal, Heikki Mannila, Ramakrishnan Srikant, Hannu Toivonen, and A. Inkeri Verkamo. Fast discovery of association rules. In Fayyad et al. [16].

[5] C. Bishop. *Neural Networks for Pattern Recognition*. Clarendon Press, 1995.

[6] Leo Breiman, Jerome H. Friedman, Richard A. Olshen, and Charles J. Stone. *Classification and Regression Trees*. Wadsworth, 1984.

[7] S. Chauduri and D. Madigan, editors. *ACM-99 Conference on Knowledge Discovery and Data Mining*, San Diego, California, 1999.

[8] Peter Cheeseman and John Stutz. *Bayesian Classification (Autoclass): Theory and Results*, pages 153 – 180. In Fayyad et al. [16], 1996.

[9] D. Chickering, D. Geiger, and D. Heckerman. Learning bayesian networks: Search methods and experimental results. In *Proceedings of the Fifth Conference on Artificial Intelligence and Statistics*, 1995.

[10] Gregory F. Cooper and Edward Herskovits. A bayesian method for the induction of probabilistic networks from data. *Machine Learning*, 9:309–347, 1992.

[11] Saul Jacka David J. Hand. *Statistics in Finance*. Arnold, 1998.

[12] Benjamin S. Duran and Patrick L. Odell. *Cluster Analysis, A Survey*. Lecture Notes in Economics and Mathematical Systems, vol 100. Springer-Verlag, 1974.

[13] R. Kohavi F. Provost, T. Fawcet. Analysis and visualization of classifier performance. *Proceedings of the 15th ICML*, 1998.

[14] Usama M. Fayyad. Branching on attribute values in decision tree generation. In *Proceedings of the 12th National Conference on Artificial Intelligence*, pages 601–606. AAAI/MIT Press, 1994.

[15] Usama M. Fayyad, Gregory Piatetsky-Shapiro, and Padhraic Smyth. *From Data Mining to Knowledge Discovery: An Overview*, pages 1 – 34. In Fayyad et al. [16], 1996.

[16] Usama M. Fayyad, Gregory Piatetsky-Shapiro, Padhraic Smyth, and Ramasamy Uthurusamy, editors. *Advances in Knowledge Discovery and Data Mining*. AAAI/MIT Press, 1996.

[17] Usama M. Fayyad and Ramasamy Uthurusamy, editors. *AAAI-95 Conference on Knowledge Discovery and Data Mining*, Montreal, Quebec, 1995.

[18] J.H. Friedman and J.W. Tukey. A projection pursuit algorithm for exploratory data analysis. *IEEE Transactions on Computing*, C-23:881–889, 1974.

[19] Jim Gray, Surajit Chaudhuri, Adam Bosworth, Andrew Layman, Don Reichart, Murali Venkatrao, Frank Pellow, and Hamid Pirahesh. Data cube: A relational aggregation operator generalizing group-by, cross-tab, and sub totals. *Data Mining and Knowledge Discovery, An International Journal*, 1, 1997.

[20] Peter Grünwald. *The Minimum description Length Principle and Reasoning under Uncertainty*. PhD thesis, University of Amsterdam, 1998.

[21] David J. Hand, Joost N. Kok, and Michael R. Berthold, editors. *Advances in Intelligent Data Analysis*, number 1642 in LNCS, Amsterdam, The Netherlands, 1999. Springer.

[22] Venky Harinarayan, Anand Rajaraman, and Jeffrey D. Ullman. Implementing data cubes efficiently. In *Proceedings of the 1996 SIGMOD Conference*, pages 205 – 216, 1996.

[23] David Heckerman, Heikki Mannila, Daryl Pregibon, and Ramasamy Uthurusamy, editors. *AAAI-97 Conference on Knowledge Discovery and Data Mining*, Newport Beach, California, 1997.

[24] John Hertz, Anders Krogh, and Richard G. Palmer. *Introduction to the Theory of Neural Networks*. Santa Fe Institute Lecture Notes vol 1. Addison-Wesley, 1991.

[25] Marcel Holsheimer, Martin Kersten, and Arno Siebes. Data surveyor: Searching the nuggets in parallel. In *Advances in Knowledge Discovery and Data Mining*, pages 447 – 467. MIT Press/AAAI Press, 1996.

[26] Peter J. Huber. Projection pursuit. *The Annals of Statistics*, 13(2):435–475, 1985.

[27] Finn V. Jensen. *An Introduction to Bayesian Networks*. Springer, 1996.

[28] Jan Komorowski and Jan Zytkow, editors. *Principles of Data Mining and Knowledge Discovery*, number 1263 in LNAI, Trondheim, Norway, 1997. Springer.

[29] John R. Koza. *Genetic programming*, volume 1. MIT Press, 1992.

[30] John R. Koza. *Genetic programming*, volume 2. MIT Press, 1994.

[31] Ming Li and Paul Vitányi. *An Introduction to Kolmogorov Complexity and its Applications*. Texts and Monographs in Computer Science. Springer Verlag, 1993.

[32] X. Liu, P. Cohen, and M. Berthold, editors. *Advances in Intelligent Data Analysis*, number 1280 in LNCS, London, UK, 1997. Springer.

[33] Hongjun Lu, Hiroshi Motoda, and Huan Liu, editors. *KDD: techniques and applications*, Singapore, 1997. World Scientific.

[34] Heikki Mannila and Kari-Jouko Räihä. Algorithms for inferring functional dependencies from relations. *Data and Knowledge Engineering*, 12:83–99, 1994.

[35] K.V. Mardia, J.T. Kent, and J.M. Bibby. *Multivariate Analysis*. Probability and Mathematical Statistics. Academic Press, 1979.

[36] D. Michie, D.J. Spiegelhalter, and C.C. Taylor, editors. *Machine Learning, Neural and Statistical Classification*. Ellis Horwood series in Artificial Intelligence. Ellis Horwood, 1994.

[37] Anthony O'Hagan. *Bayesian Inference*. Kendall's Advanced Theory of Statistics, vol 2B. Edward Arnold, 1994.

[38] S Stolfo P. Chan. Towards scalabale learning with non-uniform class and cost distributions. *Proceedings of KDD98*, 1998.

[39] J. Pearl. *Probabilistic Reasoning in Intelligent Systems*. Morgan Kaufmann, 1988.

[40] J.R. Quinlan. Induction of decision trees. *Machine Learning*, 1:81–106, 1986.

[41] J.R. Quinlan. Probabilistic decision trees. In Y. Kodratoff and R. Michalski, editors, *Machine Learning: An Artificial Intelligence Approach, Vol 3*. Morgan Kaufmann, 1990.

[42] B.D. Ripley. *Pattern Recognition and Neural Networks*. Cambridge University Press, 1996.

[43] C.P. Robert. *The Bayesian Choice*. Springer Verlag, 1994.

[44] Arno Siebes. Data surveying, foundations of an inductive query language. In Fayyad and Uthurusamy [17], pages 269–274.

[45] Evangelos Simoudis, Jiawei Han, Usama M. Fayyad, and Ramasamy Uthu- rusamy, editors. *AAAI-96 Conference on Knowledge Discovery and Data Mining*, Portland, Oregon, 1996.

[46] Alan Stuart and Keith Ord. *distribution Theory*. Kendall's Advanced Theory of Statistics, vol 1. Edward Arnold, 1994.

[47] Alan Stuart, Keith Ord, and Steven Arnold. *Classical Inference and the Linear Model*. Kendall's Advanced Theory of Statistics, vol 2A. Edward Arnold, 1999.

[48] J.W. Tukey. *Exploratory Data Analysis*. Addison-Wesley, 1977.

[49] Xindong Wu, Ramamohanarao Kotagiri, and Kevin B. Korp, editors. *Research and Development in Knowledge Discovery and Data Mining*, number 1394 in LNAI, Melbourne, Australia, 1998. Springer.

[50] Ning Zhong and Lizhu Zhou, editors. *Research and Development in Knowledge Discovery and Data Mining*, number 1574 in LNAI, Beijing, China, 1999. Springer.

[51] Jan Zytkow and Jan Rauch, editors. *Principles of Data Mining and Knowledge Discovery*, number 1704 in LNAI, Prague, Czech Republic, 1999. Springer.

[52] Jan M. Zytkow and Mohamed Quafafou, editors. *Principles of Data Mining and Knowledge Discovery*, number 1510 in LNAI, Nantes, France, 1998. Springer.

SUBGROUP MINING

W. Klösgen
German National Research Center for Information Technology (GMD), Sankt Augustin, Germany

W. Klösgen
German National Research Center for Information Technology (GMD), Sankt Augustin, Germany

ABSTRACT

Statistical findings on subgroups belong to the most popular and simple forms of knowledge we encounter in all domains of science, business, or even daily life. We read or hear such messages as: Lung cancer mortality rate has considerably increased for women during the last 10 years, unemployment rate is overproportionally high for young men with low educational level, potential of violance is the highest for males between 14 and 18. In this paper, we first compare knowledge expressed by subgroup patterns with other popular knowledge types of Knowledge Discovery in Databases (KDD), introduce types of description languages for subgroups, summarize general pattern classes for subgroup deviations and associations. A deviation pattern describes a deviating behavior of a target variable in a subgroup. Deviation patterns rely on statistical tests and thus capture knowledge about a subgroup in form of a verified (alternative) hypothesis on the distribution of a target variable. Search for deviating subgroups is organized in two phases. In a brute force search, alternative search heuristics can be applied to find a set of deviating subgroups. In a second refinement phase, redundancy elimination operators identify a system of subgroups. We discuss the role of tests for subgroup mining, introduce specializations of the general deviation pattern, summarize search approaches, and deal with navigation and visualization operations that support an analyst in interactively constructing a best system of deviating subgroups.

1. INTRODUCTION

Subgroup patterns are local findings identifying subgroups of a population with some unusual, unexpected, or deviating behavior. Thus these patterns do not give a complete overview on the population, but refer only to subsets. Nevertheless these local statements are often useful, either because the interest of the analyst is rather modest, or the available data do not allow to derive the complete description of the behavior for the whole population. In a production control application, the analyst may be satisfied, if some conditions for the large number of process variables can be identified that lead to a very high quality of the product. Then these conditions will probably be tried to steer a better production process. In a medical application, the available variables can often not describe the dependencies between symptoms and diagnoses, because only a part of the relevant and typically unknown variables is available. But it is useful to know at least some subgroups of patients that can be diagnosed with a high accuracy.

Whereas contingency tables, decision trees, and functional relations describe the complete relation between two or more variables, typically between one dependent and several independent variables, a subgroup pattern represents only some local knowledge which can specifically be seen as originating from a cell of a contingency table, a leaf of a decision tree, or a special value instantiation of a functional equation. A cluster belonging to a clustering can also be seen as a subgroup, especially when derived by conceptual clustering methods. Taxonomies and concept hierarchies are important for constructing subgroups, especially when they refer to the value domain of a single variable. Rules are special subgroup patterns that typically deal with nominal target variables.

To be useful, subgroup patterns must satisfy at least two conditions. Subgroup descriptions must be interpretable and application relevant and the reported behavior for the subgroup must be interesting which specifically means that it is statistically significant.

2. DESCRIPTION LANGUAGES FOR SUBGROUPS

A subgroup is informally defined as an interpretable and application relevant subset of a studied population. Thus the subgroup consists of objects belonging to the population. A description of the subgroup is a statement in a subgroup description language that specifies the properties that must be satisfied by the subgroup objects. The subgroup consists of those objects for which these properties are true. When the context is clear, we do not distinguish between the subgroup as a set of objects (extension) or defined by a subgroup description (intension). A subgroup description language [1] has to ensure that the subgroups constructed in this language are interpretable and application relevant. This can of course often only be partially achieved.

Subgroup descriptions rely on variables available in the data base. A *selector* is a subset of the value domain of a variable. Conjunctive description languages consist of descriptions that are built as conjunctions of selectors. Description languages for subgroup mining are mostly conjunctive, propositional languages. So we assume that the data base consists of one or several relations, each with a schema $\{A_1, A_2, ..., A_n\}$ and associated domains D_i for the variables (attributes) A_i. A conjunctive, propositional description of a subgroup is then given by:

$$A_1 \in V_1 \wedge ... \wedge A_n \in V_n \quad \text{with} \quad V_i \subseteq D_i.$$

Conjunctive selectors with $V_i = D_i$ can of course be omitted in a description.

One-relation languages allow only to analyse a single relation, possibly constructed in a preprocessing phase by a join of several relations. Then only the attributes belonging to the schema of this single relation can be used for subgroup descriptions.

Multirelational languages do not require a preprocessing join operation and allow to build descriptions with attributes from several relations. They are especially useful for applications that require different target object classes with flexible joins to be analysed. In the *Kepler* system, the *MIDOS* subgroup miner [2] discovers subgroups of objects of a selected target relation. A subgroup description may include selectors from several relations, which are linked by foreign link attributes.

A data base may, for instance, include relations on hospitals, patients, diagnoses of patients, and therapies for patient-diagnoses with obvious foreign links such as patient-id linking patient-diagnoses and patients. The analyst chooses a target object, e.g. patients, and can decide on the other object classes, e.g. hospitals and patient-diagnoses, and their attributes to be used for building subgroups of patients. Such a subgroup could e.g. be described by *male patients with a cancer diagnosis treated in small hospitals* .

By these foreign links and the implicit existential quantifier used for linking relations (e.g. patients with at least one diagnosis of a type), a very limited Inductive Logic Programming approach is applied, extending the simple one relational propositional approach. The full ILP approach has not (yet) been used for subgroup mining.

A next dimension for classifying specializations of description languages refers to the type of taxonomies that can be used for subgroup descriptions. To restrict the number of descriptions, usually not every subset of the domain of an attribute A_i is allowed in a description. A taxonomy H_i consisting of a set of subsets of D_i holds the allowed subsets. A taxonomy is hierarchically arranged, i.e. the subsets are partially ordered by inclusion. Usually H_i will be much smaller than the power set of D_i. Such a taxonomy can explicitly and statically be given for an attribute included in the description language for a mining task, or dynamically and implicitly determined by a special subsearch process that generates and evaluates certain subsets of attribute values during a mining task.

3. BEHAVIOR PATTERNS FOR SUBGROUPS

Many types of patterns can be identified for a subgroup. We distinguish two general pattern classes: deviation and association patterns. A deviation pattern describes a subgroup with some type of deviation for one (or several) designated target variables [3]. Various specializations of the general deviation pattern are treated below. An association pattern reports a type of association between two subgroups and thus identifies a pair of subgroups.

Unemployment is the designated target variable for the example given in the introduction. The subgroup description refers to *gender = male & age = young & educational level = low*, where *young* and *low* may designate some entry from a taxonomy. The deviation type is given by a target variable rate (unemployment rate) that deviates significantly from some reference value, e.g. the target rate in the whole population. An example of an association pattern would be: *Upload operations by experienced users are often followed by extensive comment operations*. When analysing logfiles of Web usage, a pair of user actions (*upload operations by experienced users* and *extensive comment operations*) is identified holding some association type. In this example, the association type is given by a mathematical relation between pairs of operations (executed in the same session, one after the other referring to time of execution) and some statistical association measure (*often*, i.e. with a high confidence rate).

Statistical significance has already been mentioned as a necessary precondition for interestingness. Subgroup patterns rely on statistical tests to decide on significance. In the null hypothesis of such a test, the uninteresting case of a not deviating subgroup is specified. The alternative hypothesis deals with the interesting deviation. Thus a subgroup pattern, from a statistical point of view, captures knowledge in the form of a verified (alternative) hypothesis.

4. MOTIVATION FOR SUBGROUP MINING AND ROLE OF HYPOTHESIS TESTING

The motivation is caused by a frequent situation where an analyst is interested in a special property or behavior of one or several selected target variables. Those regions of the input variable space are searched where the target variables show this behavior. The analyst could be interested in regions e.g. with a high average value of a continuous target, with a high share of one value of a nominal target, for which this share has significantly more increased during two years than in the complementary region, or that show a special time trend of a target variable. Thus many different data analysis questions can be represented as special behavior types of target variables.

One approach for many of these data analyses relies on an approximation of the unknown function that describes the dependency between target and input variables. The approximating function is then studied to identify the interesting regions. However, it is difficult to find a good approximation when there are many input variables, the approximation must be global (covering the whole input space), and be derived with a sample of noisy data. So often the direct approach of searching for the interesting subgroups without relying on an intermediary functional approximation is more efficient [4].

To formalize the subgroup mining approach, a specification of the description language to build subgroups and a formalization of behavior patterns are needed. The behavior of target variables $\mathbf{y} = (y_1, \dots, y_k)$ is captured by assuming a probabilistic approach and referring to their joint distribution with the input variables $\mathbf{x} = (x_1, \dots, x_m)$. One is interested in some designated property of the unknown joint distribution with density $p(\mathbf{y},\mathbf{x})$, e.g. a large mean of $E[y|\mathbf{x}]$ in a subgroup. The interesting behavior is now defined by a statistical test. In the null hypothesis of the test, the assumption on the distribution of the target variable(s) in the subgroup is specified that is regarded as expected or uninteresting. The alternative hypothesis defines the deviating or interesting subgroup. When a given data set spots the null hypothesis for a subgroup as very unlikely (under a given confidence threshold), the subgroup (i.e. the behavior of the target variable(s) in the subgroup defined by the distribution property) is identified as interesting. In this way, a broad spectrum of statistical tests and of associated data analysis questions is applicable for subgroup mining.

Thus subgroup mining does not rely on an estimation of the density function (kernel density estimation), or more simply on an estimation of the first moment $E[y|\mathbf{x}]$ of the density function (function approximation), but a direct search for significant subgroups is scheduled.

Now the problem remains to formulate appropriate null hypotheses to represent analytic questions. In principal, one can rely on a-priori assumptions on the behavior regarded as uninteresting, or on measurements within the population which in case of a given sample data base estimate the real population behavior. For instance, the probability of a binary event for members of a subgroup could be regarded as uninteresting, if it does not deviate from the overall probability for the total population. The probability for the population can be estimated with the given sample data base and used for the null hypothesis of a parametrical statistical test.

A subgroup is interesting dependent on a selected confidence threshold. Since very many tests are performed in a mining task, based on Bonferroni adjustment for multiple tests, a very high confidence threshold must be specified. Even if, due to clever heuristic search strategies, only a small part of the whole subgroup space is processed, conceptually very many tests are considered. This must be regarded when setting the confidence level, either automatically or by the user. The problem is still aggravated by secondary search processes, e.g. for discretization of continuous or value clustering of spatial independent variables.

The test approach has three advantages for subgroup mining. It allows a broad spectrum of data analytic questions to be treated, offers intelligent solutions to balance the trade-off between diverse criteria for assessing the statistical interestingness of a deviation, e.g. size of subgroup versus amount of deviation, and finally mitigates the problem of just discovering random fluctuations of the target variables in the given noisy sample as interesting.

5. SPECIALIZATIONS OF THE GENERAL DEVIATION PATTERN

Special analytic questions can be classified with two main dimensions. At first, the type of the target variable is important. For a binary target, a single share is analysed, e.g. the percentage of good productions. In case of a nominal target, a vector is studied, e.g. the joint percentages of bad, medium, and good productions. When the target is ordinal, the median, or alternatively, the probability of a better value in the subgroup than in a reference group can be analysed. E.g. the probability that a production from a subgroup is better than a production from the complementary subgroup. Finally, the target can be continuous. Then statements on the mean of the variable can be inferred. If several target variables are selected, their joint distribution must be analysed. The second dimension for classifying analytic questions is given by the number of studied populations. Populations may e.g. relate to several time points or countries.

Type of dependent variable(s)	One cross section	Two independent cross sections	k independent cross sections and time series
Binary	binomial test chi square test confidence intervals information gain	bin.test:pooled variance chi square test log odds ratio: z-scores (each with absolute relative version)	chi square tests trend test

Nominal	chi square:
	    goodness of fit
	    independence test
	Gini diversity index
	information gain
	twoing criterium
		chi square tests
		Gini diversity index
			chi square test
			trend analysis

Ordinal	ridit analysis	ridit analysis	ridits & trend analysis
Continuous	median test median-quantile test U-test H-test 1 or 2 sample t-test	median test median-quantile test U-test H-test two-sample t-test	analysis of variance

Table 1:   Some statistical verification tests for subtypes of the subgroup deviation pattern

A verification and a quality function evaluate the deviation of a subgroup. The verification test operationalizes a special analytic question. Table 1 lists some tests according to the classification of analytic questions. When relying on parametrical tests, a property of one of the distribution parameters of the target determines the meaning of the analytical question. Non-parametrical tests are appropriate in the data mining context, because the smaller test power (adhering longer to the null hypothesis) is mostly not a problem, and the modest distribution assumptions and calculation efforts of these tests are preferable. Also the large sample and explorative situation in data mining favours non-parametrical tests. The verification function is used as a filter constraint for subgroups. Only deviations are selected that have a very low probability of being generated just by random fluctuations of the targets.

The quality function is used by the search algorithm to rank the subgroups. For instance, in a beam search strategy, only the best n subgroups according to their quality are further expanded. Quality computation can rely on statistical and other interestingness aspects such as simplicity, usefulness, novelty. Quality can be given by the p-value or test statistic calculated in the verification method, or by a function exploiting this significance value as one component for the final quality. A typical statistical quality function (e.g. defined by z-scores) combines several aspects of interestingness such as *strength* (deviation of parameter from a-priori value) and *generality* (subgroup size).

Alternative tests for an analytical question can be assessed by their statistical properties (e.g. power or type II error). It is also important for deciding between several test options, how the single interestingness aspects are combined. We will discuss this, representatively for the tests listed in table 1, for the simplest case, the binary event.

$$Q_1 = \frac{p - p_0}{\sqrt{p_0(1 - p_0)}} \sqrt{n} \tag{1}$$

$$Q_2 = \frac{p - p_0}{\sqrt{p_0(1 - p_0)}} \sqrt{n} \sqrt{\frac{N}{N - n}} \tag{2}$$

Tests for a binary event in a subgroup include several criteria. The z-score quality function based on comparing the subgroup with the total population (1) balances three criteria to measure the interestingness of a subgroup: the size of the subgroup (n), the difference of the target shares ($p - p_0$), and the level of the total target share (p_0).

If the target share in the subgroup is compared with the share in its complementary subset (2), a fourth criterion is added, the relative size of the subgroup with respect to total size (N). Then large subgroups are favoured, which is useful for patient search strategies. Thus option (2) is generally more appropriate. The test options (1) and (2) combine the diverse evaluation criteria in an evident way. For the confidence interval approach (distance between intervals or p-value for touching intervals as quality), the quality formula is more complex and favours large deviations.

Option (2) has a factor $g/(1-g)$ instead of relative subgroup size g for option (1). These two factors appear also in other quality functions, e.g. in the two chi-2 tests of table 1, or the two-sample t-test. Option (2) is symmetric with the same value for a subgroup and its complementary group, which is often more appropriate (e.g. for discretization subsearch).

But the appropriate balance of criteria may depend on application requirements. Thus typically the analyst decides, based on background knowledge, which criteria are more important for a special application. Specifically, the role of subgroup size usually is application dependent. For some applications (e.g. production control), very small subgroups with large deviations are important, for other applications (e.g. marketing), subgroup size is important

6. SEARCH

Search for deviations is determined by search space dimensionality, strategies, pruning, and constraints. The main search dimension is the space of subgroup descriptions, partially ordered by the generality of the intensional descriptions or by the subset relation of their extensions. In *Explora* [3] a search algorithm for multidimensional spaces with an induced product ordering exploits additionally a space of target subgroups which are built as conjunctions of taxonomical values for target variables and a space of range subgroups. Search in this three-dimensional space is scheduled from general to specific groups regarding the product ordering and constrained by redundancy filters pruning successor subgroups. So interesting subgroups are at first identified for the whole population and the most general target groups. These patterns are then refined for more special target groups and range restrictions.

6.1 Search strategies

Search strategies usually iterate over two main steps: validating hypotheses (subgroups) and generating new hypotheses. Operating on a current population of hypotheses, neighborhood operators generate the neighbors, e.g. by expanding hypotheses with additional selectors, or genetic operators create a next generation by mutation and cross-over. Both the validation and generation step consist of four substeps summarized in table 2.

Search strategies fix the details within this general search frame, e.g. the order in which the hypotheses are evaluated, expanded and validated, the selection and pruning criteria, and the iteration, recursion or backtracking. In table 2, these steps are summarized for some simple search strategies implemented in *Data Surveyor* (Siebes, this issue).

All these strategies perform a brute force search to identify a set of hypotheses (subgroups) with high quality. Whereas beam search at each step only expands the best hypotheses to find more specialized, better subgroups, the broad view strategy is complementary. If a high quality subgroup is found, it is not further expanded. So subgroups can be identified that consist of a conjunction of selectors, where each selector alone is not interesting. The "best n" strategy is exhaustive, so that an efficient pruning is necessary for large hypothesis spaces. This can be achieved by a restrictive cover constraint (requiring a large size of a subgroup). The optimistic estimate evaluation of a subgroup checks, if any specialization can have a better quality than the worst of the currently best n hypotheses.

Another aspect of a search strategy relates to its greedyness. The usual general to specific search realized by successively adding further conjuncts is very greedy, i.e. the size of the next subgroup is much reduced by a further conjunct. Especially for hill climbing strategies, this often is a problem. Friedman and Fisher [4] therefore propose a patient strategy based on a description language offering all internal disjunctions for categorical and quantiles for continuous variables. At each specialization step, one internal disjunction is eliminated or one small upper or lower quantil is taken away from the current interval. So only a small part of the objects of a current subgroup is reduced in a specialization step.

Search Step	Beam Search	Broad View	Best n	Patient
Select hypotheses for validation from list of generated, not yet validated hypos	all	all	all	all
validate	apply verification test and quality computation			
evaluation of vali-dated hypotheses	sort successfully verified, not prunable hypotheses (cover constraint) by quality and put best n on list of hypotheses to be expanded	put not successfully verified, not prunable hypotheses (cover constraint) on list of hypotheses to be expanded put successfully verified hypos on result list	update list of best n hypos with successfully verified hypos. Put not prunable hyps (cover constraint, optimistic estimate) on list of hyps to be expanded	sort successfully verified, not prunable hypos by quality and put best one on list of hypos expanded. If no better hypo, repeat process, but eliminate or disregard all cases covered by found subgroups
update list of not yet validated hypos	not applicable: all have been validated			
select hypotheses for expansion	all	all	all	all
expand hypos	dep. on type of expansion attribute: discretization, regional clustering			eliminate 1 intern. disjunction, resp. 1 quantile
evaluation of expanded hypos		eliminate successors of results		
update list of hypos to be expanded	not applicable: all have been expanded			

Table 2: Four simple brute force search strategies for subgroup mining

6.2 Search constraints

Pruning criteria are special constraints limiting the search space and thus are important to realize an efficient algorithm. The main pruning criterium for subgroup mining is the cover constraint. Other criteria are given by syntactical details of the description language, e.g. maximal number of conjunctions. Depending on the search strategy, quality estimates can be exploited. When e.g. searching for the best n subgroups, all those subgroups can be pruned for which all successors that fulfill the cover constraint cannot have a higher quality than the current n-th best subgroup [5]. These optimistic estimates depend on the quality function, so that for each function an estimate must be derived. In [3] some general classes of quality functions are analysed, for which an easy calculation of such an estimate is possible. However, these estimates show some different power, i.e. the pruning potential for the diverse quality functions is very different.

6.3 Search refinement

In a brute force phase, subgroups are determined satisfying constraints and goals of the selected search strategy. In a refinement phase, subgroups are elaborated, redundancies eliminated, and general patterns specialized. Elaborations treat a single hypothesis by filtering (during and after search), bottom up refinement, sensitivity analysis of description boundaries, statistical pruning and cross validation, or analysing the homogeneity of subgroups to avoid that not the subgroup as a whole is relevant but a subset. Redundancies relate to the correlation between subgroups which may include spurious effects. Brute force and refinement subtasks can be scheduled iteratively. This can be done automatically or in an user controlled exploratory mode. Redundancies are eliminated by suppressing and ordering, combining, generalizing or clustering of hypotheses [3], [6].

The overall goal is to find a consolidated set of subgroups. Thus criteria are needed to evaluate a set of subgroups. A selection process selects or produces (by constructing additional subgroups) a best set of subgroups. Specifically, four criteria are important: The overlapping degree of the selected subgroups should be low. The covering degree of subgroups should be high. If e.g. the target objects are determined on an individual level, the covering degree is given by the percentage of target objects that are included in the union of the selected subgroups. The quality of the union of the selected subgroups (regarded as a single subset) should be high. And finally, the number of selected subgroups should be low. Trade-offs between these criteria can be automatically scheduled by thresholding parameters. In contrast to an automatic refinement, user involvement in this process is however usually more appropriate.

7. NAVIGATION AND VISUALIZATION

The kind and extent of user involvement into a data mining step considerably variates dependent on applications and user preferences. Subgroup mining systems differ in the degree of autonomy that is incorporated in the system by the parameterization of decision processes and treatment of trade-offs between evaluations aspects. Because the autonomy requirements of an analyst are so different, a system should provide both a nearly fully automatic search and refinement identifying a consolidated best set of pattern instances for a mining task, and a user controlled iterative and explorative search for such a set of patterns. A user centered search is incrementally scheduled and supported by navigation operators to specify and redefine search tasks to be run in subspaces of a hypothesis space.

Visualization of search results is important for these navigation operators. The analyst should be able to operate on the presented results to perform comparison, focusing, explanation, browsing and scheduling operations. Visualization deals with four aspects: how can a single pattern and a set of interdependent patterns be appropriately visualized, which interactive operations can be performed on a presentation graph, and which additional visualizations are important to explain the results and support explorative analyses.

Patterns and sets of patterns must be presented in textual and graphical form. A set of patterns can often be represented as a graph referring to the partial ordering by generality. Various operations on these graphs allow the user of a mining system to redirect a mining task, to filter or group mining results, and to browse into the data base. Thus these graphs provide an interaction medium for the user based on interactive visualization techniques.

Besides the interactivity of operations on the visualized subgroup mining results, the pattern specific presentation of a single subgroup must be designed. Text presentations of subgroups can be simply arranged with presentation templates. Additionally, appropriate graphical presentations of subgroups and their deviation figures must be determined. For example, a simple share pattern (binary dependent variable, one population) can be graphically represented as a fourfold display including the confidence intervals of the share [7]. In case of a nominal dependent variable, the set of percentages could be represented as a pie chart. However, already the application of pie charts to illustrate a single frequency distribution is heavily discussed, because of the limited capacities of humans to compare a set of angles. Using many pie charts to compare several subgroups and their frequency distributions for the values of a nominal variable is even more doubtful.

Additional visualizations to explain the mining results can e.g. support the analyst to select between subgroups by assessing the trade-off between generality and strength. Friedman and Fisher [4] propose a trajectory visualization of subgroups in a two dimensional generality vs. strength space. Another example relates to the multicollinearity problem. A frequency distribution of the values of an input variable for a subgroup and the population can help to identify those correlations. Other visualizations can uncover the overlapping degree between subgroups and explain a suppression refinement. Friedman and Fisher also propose sensitivity plots that can be used to judge the sensitivity of the hypothesis (subgroup) quality to the subgroup description boundaries. With these plots, the overfitting problem is addressed.

CONCLUSION

The generic components of deviation patterns include a description language to construct subgroups, a verification method to test the significance of a subgroup, quality functions to measure the interestingness of a single subgroup and of a set of subgroups, constraints limiting the space of admissible subgroups, and search goals and controls defining additional properties of the subgroups to be found. Interactive visualization of individual subgroups and sets of interdependent subgroups is fixed in the presentation component of a pattern class. Subgroup mining is a pragmatic exploration approach that can be applied for various analytic questions. Althogh subgroup mining has reached a quite impressive development status, it is an evolving area for which a lot of problems must still be solved which are finally summarized.

Assessing the validity of discovery results must be elaborated for subgroup patterns, e.g. to avoid that they overfit the given data. A second problem relates to providing adequate description languages, such as multi relational languages and dynamic, constructive induction of additional variables that are better suited to describe the given data. Especially for time and space related data, such derived variables can be useful when including descriptive terms e.g. based on means, slopes or other time series indicators. Finding a best set of hypotheses among a large set of significant hypotheses and integrating several aspects of interestingness, e.g. significance, novelty, simplicity, usefulness, are next problems. Robustness means that discovery results should not differ too sensitively respective to small alterations of the data, description language or selected values of the dependent variables. The main concern in KDD has been on accuracy, whereas robustness until now only plays a minor role in discovery research. *Second order discovery* to compare and combine the results for different pattern types could be necessary, especially if many analysis questions are issued to the data. A next point relates to changing data and domain knowledge. This problem area includes incremental mining methods adapting existing results according to the addition or modification of a small number of tuples and comparing new discovery results with the preceding results. Finally there are a lot of technical challenges to ensure efficient and interactive KDD processes. High performance solutions are necessary for VLDB applications. Other problems relate to the integration of KDD systems with other systems such as database systems or statistical packages.

REFERENCES

1. Michalski, R.S.: A Theory and Methodology of Inductive Learning, in: Machine Learning: An Artificial Intelligence Approach (eds. Michalski, R.S.; Carbonell, J. and Mitchell, T.), Tioga Publishing, Palo Alto 1983, 83–134.
2. Wrobel, S.: An Algorithm for Multi-relational Discovery of Subgroups, in: Proceedings of the First European Symposium on Principles of KDD (eds. Komorowski, J. and Zytkow, J.), Springer-Verlag, Berlin 1997, 78–87.
3. Klösgen, W.: Explora: A Multipattern and Multistrategy Discovery Assistant, in: Advances in Knowledge Discovery and Data Mining (eds. Fayyad, U.; Piatetsky-Shapiro, G.; Smyth, P. and Uthurusamy, R.), MIT Press, Cambridge 1996. 249–271.
4. Friedman, J. and Fisher, N.: Bump Hunting in High-Dimensional Data, in: Statistics and Computing 1998.
5. Smyth, P. and Goodman, R.: An information theoretic approach to rule induction, in: IEEE Trans. Knowledge and Data Engineering 4, 1992.
6. Gebhardt, F.: Choosing among Competing Generalizations, in: Knowledge Acquisition 3, 1991.
7. Friendly, M.: Conceptual and Visual Models for Categorical Data, in: The American Statistician 1993.

POSSIBILISTIC GRAPHICAL MODELS

C. Borgelt
Otto-von-Guericke University, Magdeburg, Germany

J. Gebhardt
TU Braunsweig, Braunsweig, Germany

R. Kruse
Otto-von-Guericke University, Magdeburg, Germany

Abstract

Graphical modeling is an important method to efficiently represent and analyze uncertain information in knowledge-based systems. Its most prominent representatives are Bayesian networks and Markov networks for probabilistic reasoning, which have been well-known for over ten years now. However, they suffer from certain deficiencies, if imprecise information has to be taken into account. Therefore *possibilistic graphical modeling* has recently emerged as a promising new area of research. Possibilistic networks are a noteworthy alternative to probabilistic networks whenever it is necessary to model both uncertainty *and* imprecision. Imprecision, understood as set-valued data, has often to be considered in situations in which information is obtained from human observers or imprecise measuring instruments. In this paper we provide an overview on the state of the art of possibilistic networks w.r.t. to propagation and learning algorithms.

1 Introduction

A major requirement concerning the acquisition, representation, and analysis of information in knowledge-based systems is to develop an appropriate formal and semantic framework for the effective treatment of uncertain and imprecise data [32]. In this paper we consider this requirement w.r.t. a task that frequently occurs in applications, namely the task to identify the true state ω_0 of a given world section. We assume that possible states of the domain under consideration can be described by stating the values of a finite set of attributes (or variables). The set of all possible (descriptions of) states, i.e., the Cartesian product of the attribute domains, we call the *frame of discernment* Ω (also called *universe of discourse*). The task to identify the true state consists in combining *generic knowledge* about the relations between the values of the different attributes (usually derived from background expert knowledge about the domain or from databases of sample cases) and *evidential knowledge* about the current values of some of the attributes (obtained, for instance, from observations). The goal is to find a description of the true state ω_0 that is as specific as possible.

As an example consider medical diagnosis. Here the true state ω_0 is the current state of health of a given patient. All possible states can be characterized by attributes describing properties of patients (like sex or age) or symptoms (like fever or high blood pressure) or the presentness or absence of diseases. The generic knowledge consists in a model of the medical competence of a physician, who knows about the relations between symptoms and diseases in the context of other properties of the patient. It may be gathered from medical textbooks or reports. The evidential knowledge is obtained from medical examination and answers given by the patient, which, for example, reveal that she is 42 years old and has 39° fever. The goal is to derive a full description of her state of health in order to determine which disease or diseases are present.

Imprecision, understood as set-valued data, enters our considerations due to two reasons. In the first place, generic knowledge about dependences between attributes can be relational rather than functional, so that knowing exact values for the observed attributes does not allow us to infer exact values for the other attributes, but only sets of possible values. Secondly, the available information about the observed attributes can itself be imprecise. That is, it may not enable us to fix a specific value, but only a set of alternatives. In such situations we only know for sure that the current state ω_0 lies within a set of alternative states, but we may have no preferences that could help us to single out the true state ω_0 from this set. For example, in medical diagnosis a physician may consider a set of diseases, all of which can explain the observed symptoms and one of which must be the correct diagnosis, without preferring any of them.

Uncertainty arises from the fact that often the functional or relational dependences between the involved attributes are unreliable or, in general, indeterministic. This situation, of course, could also be modeled as imprecision. However, often additional information is available that allows us to state preferences between the possible alternatives. If, for example, the symptom *fever* is observed, then various disorders may be the cause of this symptom. But in the absence of other information a physician will prefer a severe cold as a diagnosis, since it is a fairly common disorder. The preferences assigned to the alternatives can be quantified, for example, by degrees of confidence. They are modeled in an adequate calculus, e.g., using probability theory or possibility theory or any other non-standard uncertainty calculus. Alternatively they can be handled in a purely qualitative way by fixing a reasonable preference relation.

In the following discussion, for simplicity, we restrict ourselves to attributes with *finite* domains. We assume that the generic knowledge models prior information about the uncertainty of the truth of propositions $\omega = \omega_0$ for all alternatives $\omega \in \Omega$. Such knowledge can often be formalized as a *distribution function* on Ω, for example, as a probability distribution, a mass distribution, or a possibility distribution, depending on the uncertainty calculus that best reflects the structure and the contents of the given knowledge. Evidential knowledge about ω_0 is taken into account by *conditioning* the available generic knowledge, that is, by conditioning a given prior distribution on Ω. This process is usually based on instantiations of particular variables. In our medical example, for instance, the variable *fever* can be instantiated by measuring the body temperature of the patient. Such instantiations give rise to an inference process that computes the posterior marginal distributions for the uninstantiated variables.

Since in applications the number of attributes to be considered is usually fairly large and the size of the frame of discernment Ω grows exponentially with the number of attributes, the reasoning process described above tends to be intractable in the domain as a whole. To make reasoning feasible, knowledge representation methods take advantage of *independences* between the attributes under consideration. Such independences allow us to decompose the generic knowledge represented by the prior distribution on Ω into distributions on lower-dimensional subspaces. An important method to represent the resulting decomposition is *graphical modeling*. It also provides useful theoretical and practical concepts for efficient reasoning under uncertainty [54, 6, 35, 48]. Applications of graphical models can be found in a large variety of areas including diagnostics, expert systems, planning, data analysis, and control. For an overview, see [8].

In this paper, we focus on graphical modeling with *possibility theory* as the underlying uncertainty calculus. In section 2 we review the basics of graphical modeling and in section 3 we outline evidence propagation in graphical models. Theoretical underpinnings of *possibilistic graphical models* including the fundamental concepts of *possibilistic conditional independence*, *conditional independence graphs*, *decomposition*, and *factorization* are presented in section 4. In section 5, we discuss a specific data mining problem, namely how to *induce possibilistic graphical models* from databases of sample cases. Finally, section 6 we draw conclusions from our discussion.

2   Graphical Models

A *graphical model* consists of a qualitative and a quantitative component. The *qualitative (or structural) component* is a *graph* (hence the name graphical model), for example, a directed acyclic graph (DAG), an undirected graph (UG) or a chain graph (CG). Each node of this graph represents an attribute and each edge a direct dependence between two attributes. The structure of the graph encodes in a specific way the conditional independences between the attributes. Therefore it is often called a *conditional independence graph*.

The *quantitative component* of a graphical model is a family of distribution functions on subspaces of Ω. For which subspaces distribution functions have to be specified is determined by the structure of the conditional independence graph. If it is a directed acyclic graph, there is one (conditional) distribution function for each attribute and each possible instantiation of its parents (i.e., its predecessors in the graph), for example, a conditional probability distribution. In this case each distribution function represents the uncertainty about the value of an attribute given a specific instantiation of its parents. If the conditional independence graph is an undirected graph, there is one (marginal) distribution function, for instance, for each maximal clique of the graph, where a clique is a fully connected subgraph, and it is maximal, if it is not contained in

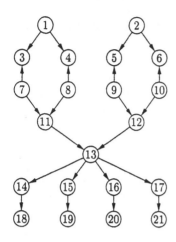

21 attributes:
1 – dam correct?	11 – offspring ph.gr. 1
2 – sire correct?	12 – offspring ph.gr. 2
3 – stated dam ph.gr. 1	13 – offspring genotype
4 – stated dam ph.gr. 2	14 – factor 40
5 – stated sire ph.gr. 1	15 – factor 41
6 – stated sire ph.gr. 2	16 – factor 42
7 – true dam ph.gr. 1	17 – factor 43
8 – true dam ph.gr. 2	18 – lysis 40
9 – true sire ph.gr. 1	19 – lysis 41
10 – true sire ph.gr. 2	20 – lysis 42
	21 – lysis 43

The grey nodes correspond to observable attributes.

Figure 1: Conditional independence graph of a graphical model for genotype determination and parentage verification of Danish Jersey cattle in the F-blood group system.

another clique. In this case each distribution function represents the uncertainty about the values of the projections of ω_0 onto the subspace corresponding to the maximal clique (i.e., the Cartesian product of the domains of the attributes contained in the maximal clique).

As an example we consider an application of a graphical model for blood group determination of Danish Jersey cattle in the F-blood group system, whose primary purpose is parentage verification for pedigree registration [37]. The underlying domain is described by 21 attributes, eight of which are observable. The size of the domains of these attributes ranges from two to eight possible values. The total frame of discernment has $2^6 \cdot 3^{10} \cdot 6 \cdot 8^4 = 92\,876\,046\,336$ possible states. Therefore a decomposition of the expert knowledge about this domain is clearly necessary to make reasoning feasible. Figure 1 lists the attributes and shows the conditional independence graph of this graphical model, which was designed by human domain experts (the graphical model is a Bayesian network and thus the conditional independence graph is a directed acyclic graph). The grey nodes correspond to the observable attributes.

The conditional independence graph reflects, as already said above, the conditional independences between the attributes of the underlying domain. In the case of a directed acyclic graph they can be read from the graph using a graph theoretic criterion called *d-separation* [35, 24]. What is to be understood by *conditional independence* depends on the uncertainty calculus the graphical model is based on. In the example at hand, which is a probabilistic graphical model, it means conditional stochastic independence of the random variables that are represented by the nodes of the graph. The joint probability distribution of these random variables is supposed to satisfy all independence relations represented by the conditional independence graph. Therefore, the joint probability distribution can be decomposed into a product of conditional

sire correct	phenogroup 1 true sire	stated sire phenogroup 1 F1	V1	V2
yes	F1	1	0	0
yes	V1	0	1	0
yes	V2	0	0	1
no	F1	0.58	0.10	0.32
no	V1	0.58	0.10	0.32
no	V2	0.58	0.10	0.32

Table 1: Conditional probability distributions for a subgraph of the conditional independence graph shown in figure 1.

probability distributions (this is also called *factorization*). This product can easily be read from the conditional independence graph: There is exactly one factor for each attribute, which refers to the conditional probability distribution of the values of this attributes given an instantiation of the parents of this attribute [35, 54].

In the Danish Jersey cattle example, a decomposition of the joint probability distribution according to the conditional independence graph shown in figure 1 leads to a considerable simplification. Instead of having to determine the probability of each of the 92 876 046 336 elements of the 21-dimensional frame of discernment Ω, only 306 conditional probabilities in subspaces of at most three dimensions need to be specified. An example of a conditional probability table is shown in table 1, which is for the phenogroup 1 of the stated sire of a given calf conditioned on the phenogroup of the true sire of the calf and whether the sire was correctly identified. The numbers in this table are derived from statistical data and the experience of domain experts. The family of all 21 conditional probability tables forms the quantitative part of the graphical model for the Danish Jersey cattle example.

3   Evidence Propagation

After a graphical model has been constructed, it can be used to do reasoning. In the Danish Jersey cattle example, for instance, the phenogroups of the stated dam and the stated sire can be determined and the lysis values of the calf can be measured. From this information the probable genotype of the calf can be inferred and it is thus possible to assess whether the stated parents of the calf are the true parents.

However, reasoning in a graphical model is not always completely straightforward. Considerations of efficiency make it often advisable to transform a graphical model into a form that is better suited for propagating the evidential knowledge and computing the resulting marginal distributions for the unobserved attributes. We briefly sketch here a popular efficient reasoning method known as *clique tree propagation* (CTP) [33, 8], which involves transforming the conditional independence graph into a clique tree.

This transformation is carried out as follows: If the conditional independence graph is a directed acyclic graph, it is first turned into an undirected graph by constructing its associated *moral graph* [33]. A moral graph is constructed from a directed acyclic

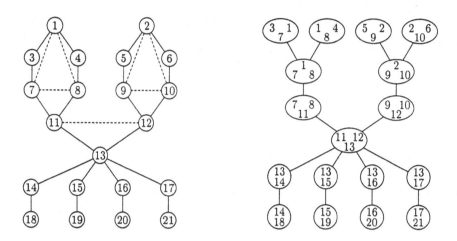

Figure 2: Triangulated moral graph (left) and clique tree (right) for the graphical model shown in Figure 1. The dotted lines are the edges added when parents were "married".

graph by "marrying" the parent nodes of all nodes (hence the name "moral graph"). This is done by simply adding undirected edges between the parents. The directions of all other edges are discarded. In general the moral graph satisfies a only subset of the independence relations of the underlying directed acyclic graph, so that this transformation may result in a loss of independence information. The moral graph for the Danish Jersey Cattle example is shown on the left in figure 2. The edges that were added when parents were "married" are indicated by dotted lines.

In a second step, the undirected graph is triangulated. (If the conditional independence graph is an undirected graph right from the start, this is the first step to be carried out.) An undirected graph is called *triangulated*, if all cycles containing at least four nodes have a chord, where a chord is an edge that connects two non-adjacent nodes of the cycle. To achieve triangulation, it may be necessary to add edges, which may result in a (further) loss of independence information. In the Danish Jersey cattle example, however, the moral graph shown on the left in figure 2 is triangulated right away, so no new edges need to be introduced.

Finally, the triangulated graph is turned into a clique tree by finding the maximal cliques, where a clique (see above) is a fully connected subgraph, and it is maximal, if it is not contained in another clique. In the clique tree there is one node for each maximal clique of the triangulated graph and its edges connect nodes that represent cliques having attributes in common. It should be noted that in general the clique tree is not unique, because often different sets of edges can be chosen. The clique tree for the Danish Jersey cattle example is shown on the right in figure 2. Detailed information on triangulation, clique tree construction and other related graph-theoretical problems can be found in [8].

The quantitative part of a graphical model, of course, has to be transformed, too. From the quantitative information of the original graphical model one has to compute a marginal distribution for each of the subspaces represented by the nodes of the clique tree. For the Danish Jersey cattle example, we have to compute a marginal distribution for the subspace formed by the attributes 1, 3, 7, one for the subspace formed by the attributes 1, 4, 8, and so on. That an appropriate factorization of the probability distribution can be found is ensured by the *Hammersley-Clifford theorem* [29]. It establishes a correspondence between the Markov properties (local, pairwise and global) of a strictly positive probability distribution P on Ω that are represented by a conditional independence graph and the factorization of P into a product of functions that depend only on the variables in the maximal cliques of the conditional independence graph.

Having constructed a clique tree, which is merely a preliminary operation to make evidence propagation more efficient, we can finally turn to evidence propagation itself. Evidence propagation in clique trees is basically an iterative extension and projection process. When evidence about the value of an attribute becomes available, it is first extended to a clique tree node the attribute is contained in. This is done by conditioning the associated marginal distribution. We call this an extension, since by this conditioning process we go from restrictions on the values of a single attribute to restrictions on tuples of attribute values. Hence the information is extended from a single attribute to a subspace formed by several attributes. Then the conditioned distribution is projected to all intersections of the clique tree node with other nodes. Via these projections the information can be transferred to other nodes, where the process repeats: First it is extended to the subspace represented by the node, then it is projected to the intersections connecting it to other nodes. The process stops when all nodes have been updated.

The propagation scheme outlined above and the subsequent computation of posterior marginal distributions for the unobserved attributes can easily be implemented by locally communicating node- and edge-processors. These processor also serve the task to let pieces of information "pass" each other without interaction. Such bypassing is necessary, if the propagation operations in the underlying uncertainty calculus are not idempotent, that is, if incorporating the same information twice can invalidate the results. This is the case, for example, in probabilistic reasoning. This problem is also the reason why the clique graph is usually required to be a *tree*: If there were loops, information could travel on two or more different paths to the same destination and thus be incorporated twice. Some other calculi, for instance, possibility theory, do not suffer from this inconvenience, so that the node- and edge-processor can be made simpler and loops do no harm (although they can make reasoning less efficient).

A well-known interactive software tool for probabilistic reasoning in clique trees is HUGIN [1]. A similar approach was implemented in POSSINFER [22] for the possibilistic setting. That the propagation is efficient is obvious: If all available evidence is entered at the same time, distributing the information in the network requires only two traversals of the clique tree.

Of course, clique tree propagation is not the only possible propagation scheme. Others include bucket elimination [11, 57] and iterative proportional fitting [54]. Commonly used propagation algorithms differ from each other w.r.t. the network structures they support, but in most cases they are applicable independent of the given uncertainty calculus, provided, of course, the elementary operations like extension (conditioning) and projection have been adapted to this calculus [33, 35, 44]. A fairly general approach to reasoning under uncertainty in so-called *valuation-based networks* has been proposed in [46, 48, 47]. It can be applied, for example, to upper and lower probabilities [52], Dempster-Shafer theory of evidence [12, 13, 42, 43, 49], and possibility theory [56, 15, 16], and has been implemented in the software tool PULCINELLA [41].

4   Possibilistic Networks

Our review of graphical models in the preceding sections was strongly oriented at the well-known theory of probabilistic networks. In this section we turn to possibilistic networks, which are a much younger but very promising type of graphical models that can deal with uncertainty *and* imprecision. Their theory can be developed in close analogy to the probabilistic case. Technically, a *possibilistic network* is a graphical model whose quantitative component is a family of possibility distributions. Hence we start our discussion by briefly recalling possibility theory and its interpretations.

Axiomatically, a *possibility distribution* π is a mapping from a reference set Ω to the unit interval. In contrast to a probability distribution, which is also defined to be such a mapping, a possibility distribution need not be normalized to one. That is, the sum (or the integral) of the *degrees of possibility* assigned by π to the elements of Ω need not be one. In the context of graphical models, we use a possibility distribution to specify imperfectly the current state ω_0 of the domain under consideration. From an intuitive point of view, $\pi(\omega)$ quantifies the possibility that the proposition $\omega = \omega_0$ is true: $\pi(\omega) = 0$ means that $\omega = \omega_0$ is impossible, whereas $\pi(\omega) = 1$ means that it is possible without restriction. Any intermediary possibility degree $\pi(\omega) \in (0, 1)$ indicates that $\omega = \omega_0$ is possible only with restrictions. That is, there is evidence supporting this proposition as well as evidence contradicting it.

Of course, the above intuitive description is much too vague to fix a particular interpretation of degrees of possibility. Thus, similar to the probabilistic case where logical, empirical, and subjective interpretations of probability can be distinguished, there is a large variety of suggestions for semantics of a degree of possibility. Among them are the view of possibility distributions as epistemic interpretations of fuzzy sets [56], the axiomatic approach to possibility theory based on possibility measures [15, 16], and the approach that bases possibility theory on likelihoods [14]. In connection to Dempster-Shafer theory, possibility distributions are seen as contour functions of consonant belief functions [42], and in the framework of set-valued statistics, they are interpreted as falling shadows [53]. Furthermore, there are interpretations of possibility theory that owe nothing to probability theory. We mention here the interpretation of possibility as *similarity*, which is related to metric spaces [39, 38, 40], and possibility as *preference*, which is justified mathematically by comparable possibility relations [17].

If one introduces possibility distributions as information-compressed representations of databases of (possibly imprecise) sample cases, as we will do in the next section, it is convenient to interpret them as (non-normalized) one-point coverages of random sets [34, 27]. This interpretation leads to very promising semantics [19, 20]. For instance, with this approach it is quite simple to establish Zadeh's *extension principle* [55] as the appropriate way of extending operations on sets to operations on possibility distributions. It turns out that the extension principle is the only way of operating on possibility distributions that is consistent with this semantic background [18].

More precisely, let Ω be the set of all possible states of the world, $\omega_0 \in \Omega$ the current (but unknown) state of the world, $(C, 2^C, P)$, $C = \{c_1, \ldots, c_k\}$, a finite probability space, and $\gamma : C \to 2^\Omega$ a set-valued mapping. C is seen as a set of contexts that have to be distinguished for a imprecise (set-valued) specification of ω_0. The contexts are supposed to describe different physical and observation-related frame conditions. $P(\{c\})$ is the (subjective) probability of the (occurrence or selection of the) context c.

A set $\gamma(c)$ is assumed to be the *most specific correct set-valued specification* of ω_0, which is implied by the frame conditions that characterize the context c. By "most specific set-valued specification" we mean that $\omega_0 \in \gamma(c)$ is guaranteed to be true for $\gamma(c)$, but is not guaranteed for any proper subset of $\gamma(c)$. The resulting *random set* $\Gamma = (\gamma, P)$ is an imperfect (i.e. imprecise *and* uncertain) specification of ω_0. Let π_Γ denote the *one-point coverage of* Γ (the *possibility distribution induced by* Γ), which is defined as

$$\pi_\Gamma : \Omega \to [0, 1], \quad \pi_\Gamma(\omega) \mapsto P(\{c \in C \mid \omega \in \gamma(c)\}).$$

In a complete modeling, the contexts in C must be specified in detail, so that the relationships between all contexts c_j and their corresponding specifications $\gamma(c_j)$ are made explicit. But if the contexts are unknown or ignored, then $\pi_\Gamma(\omega)$ is the total mass of all contexts c that provide a specification $\gamma(c)$ in which ω_0 is contained, and this quantifies the *possibility of truth* of the statement "$\omega = \omega_0$" [19, 21].

As emphasized above, graphical models take advantage of conditional independence relations in order to reduce reasoning to operations on distributions on low-dimensional subspaces. Therefore a theoretical investigation of possibilistic graphical models has to start with the definition of an appropriate concept of *conditional possibilistic independence*. Such a definition allows us to introduce *conditional independence graphs* and to search for appropriate *decomposition* and *factorization* techniques. However, in contrast to the notion of probabilistic conditional independence, which has been well-known for a long time, there is still some discussion going on about an analogous concept for the possibilistic setting. The main reason for this is the fact that with possibility theory one can model two different kinds of imperfect knowledge: uncertainty and imprecision. Hence there are at least two alternative ways of approaching the task to define conditional possibilistic independence. In addition, different semantics for possibility distributions may call for different concepts of conditional independence. For an overview, see [7].

Nevertheless, all suggestions for a concept of possibilistic conditional independence agree to the following general description: Let X, Y, and Z be three disjoint sets of attributes, X and Y non-empty. X is called *independent* of Y given Z w.r.t. a possibility distribution π on Ω, if for all instantiations of the attributes in Z, no information about the values of the attributes in Y changes the possibility degrees of the tuples over attributes in X. In other words: If the Z-values of ω_0 are known, but arbitrary, then from additional information about the Y-values of ω_0 no restrictions on the X-values of ω_0 can be derived. In terms of projecting and conditioning possibility distributions we can rephrase this concept as follows: Suppose that a possibility distribution π is used to specify imperfectly the state ω_0. If crisp knowledge about the Z-values of ω_0 is given, this distribution is conditioned w.r.t. the instantiations of the attributes in Z. If X and Y are independent given Z, then projecting the resulting conditional possibility distribution π directly to the attributes in X leads to the same distribution as first conditioning it first w.r.t. an arbitrary instantiation of the attributes in Y and only afterwards projecting it to the attributes in X.

How *conditioning* and *projection* have to be defined depends on the chosen semantics of possibility distributions: If we view possibility theory as a special case of *Dempster-Shafer theory* by interpreting a possibility distribution as a representation of a consonant belief function or of a nested random set, then the concept of conditional independence can be derived from so-called *Dempster conditioning* [42]. If we see possibility distributions as (non-normalized) one-point coverages of random sets, we have to choose the conditioning and the projection operation in conformity with the extension principle. The resulting concept of conditional independence is *conditional possibilistic non-interactivity* [28]. For details, see [7]. It should be pointed out that both types of conditional independence mentioned above satisfy the *semi-graphoid axioms* which have been established as basic requirements for any reasonable concept of conditional independence in graphical models [35]. Possibilistic conditional independence derived from Dempster conditioning even satisfies the *graphoid axioms* [36], just as probabilistic conditional independence does.

If we confine ourselves to conditional possibilistic non-interactivity in accordance with the interpretation of possibility distributions we preferred above, it is straightforward to define conditional possibilistic independence graphs: An undirected graph is called a *conditional independence graph* of a possibility distribution π, if for any three disjoint sets X, Y, and Z of nodes, X and Y non-empty, X is independent of Y given Z, if X and Y are separated by Z, i.e., if all paths from a node in X to a node in Y contain a node in Z. This definition assumes that the so-called *global Markov property* holds for π [54]. In contrast to probability distributions, where the equivalence of the global, local, and pairwise Markov property can be proven, in the possibilistic setting we have to rely on the global Markov property as the strongest of the three [18].

A proof of a possibilistic counterpart of the *Hammersley-Clifford theorem* [29] (see

lysis 40	lysis 41	lysis 42	lysis 43	genotype offspring
$\{0, 1, 2\}$	6	0	6	$V2/V2$
0	5	4	5	$\{V1/V2, V2/V2\}$
2	6	0	6	*
5	5	0	0	$F1/F1$

Table 2: A small database with four sample cases

above) is given in [18]: A possibility distribution π on Ω has a decomposition into complete irreducible components, if it is decomposed w.r.t. a triangulated conditional independence graph G of π. The *factorization* of π w.r.t. this decomposition uses the minimum instead of the product, which is used in the probabilistic case, and the maximum instead of the sum. That is, π can be represented as the minimum of its maximum projections to the maximal cliques of G. It follows that evidence is propagated in possibilistic networks with a minimum/maximum scheme instead of the product/sum scheme of the probabilistic case.

5 Learning Possibilistic Networks from Data

A graphical model is a powerful tool to do reasoning—as soon as it is constructed. Its construction by human experts, however, can be tedious and time consuming. Therefore recent research in probabilistic as well as in possibilistic graphical models focused on learning them from a database of sample cases. In accordance with the two components of graphical models, one distinguishes between *quantitative network induction*, which serves to estimate the distribution functions of the factorization represented by a graphical model, and *qualitative* or *structural network induction*, which serves to find a conditional independence graph that captures (most) of the independences of the distribution function that is induced by the database. In possibilistic learning a special concern is to exploit the information contained in imprecise, i.e., set-valued, sample cases, which pose problems for probabilistic approaches.

We start our discussion by showing how a database of imprecise sample cases induces a possibility distribution in the interpretation outlined in the preceding section. To this end we reconsider the Danish Jersey cattle example by looking at a small section of a database for this example as shown in figure 2. For simplicity, this database is reduced to five attributes. Each tuple describes one sample case, i.e., one calf. The first three tuples are imprecise, the fourth tuple is precise. The first tuple, for instance, represents the three precise tuples $(0, 6, 0, 6, V2/V2)$, $(1, 6, 0, 6, V2/V2)$, and $(2, 6, 0, 6, V2/V2)$. This means that three states have to be regarded as possible alternatives. Analogously, the second tuple represents two alternatives which result from the imprecision in the attribute *genotype offspring*. The third tuple is imprecise, because it contains an unknown value, indicated by a star '*', which can be interpreted as representing the whole domain of the corresponding attribute.

To induce a possibility distribution from this database, we interpret each tuple as corresponding to one context (see the preceding section). Assuming that the four sample cases are equally representative, it is reasonable to fix their probability of occurrence to 1/4. Note, however, that this is not enough for the probabilistic case, since it does not allow us to assign probabilities to the elementary events, i.e., the precise tuples in the domain underlying table 2. In the probabilistic setting, we may apply the *insufficient reason principle*, which states that alternatives in set-valued sample cases are equally likely, if no preferences are known. Assuming uniform distributions on set-valued sample cases leads to a refined database of $3 + 2 + 6 + 1 = 12$ precise tuples, in which, for instance, $(2, 6, 0, 6, V2/V2)$ has a probability of $1/3 * 1/4 + 0 + 1/6 * 1/4 + 0 = 3/24$. This approach, however, unjustifiably introduces information about the relative probability of the possible values. In a possibilistic interpretation of the database, we obtain for the same tuple a degree of possibility of $1/4 + 0 + 1/4 + 0 = 1/2$, since this tuple is considered to be possible in the first and in the third sample, but it is excluded in the other two. That is, no information is introduced that is not contained in the database. If we compute the possibility degrees for all tuples of the joint domain of the five attributes used in table 2, we arrive at an information-compressed interpretation of the database in the form of a possibility distribution.

Quantitative Network Induction. Whereas quantitative network induction for both probabilistic and possibilistic networks is a rather trivial task, if all sample cases are precise (standard statistical techniques can be used in this case, see [50] for an overview), sample cases with missing values and especially with imprecise (set-valued) information pose a problem. This is true even for a possibilistic approach, which is better suited to handle set-valued information, since the imprecise tuples can "overlap", thus preventing us from using simple techniques to compute maximum projections directly from the database. Fortunately, however, the database to learn from can be transformed by computing its *closure under tuple intersection*. From the transformed database all projections can be computed as efficiently as in the probabilistic case [5].

Qualitative Network Induction. The task to find a decomposition of the possibility distribution induced by a database of sample cases that best approximates this distribution w.r.t. a chosen class of conditional independence graphs is NP-hard for non-trivial classes of graphical models. This is true even if we confine ourselves to n-ary relations, which can be regarded as special cases of n-dimensional possibility distributions. For this reason, in analogy to learning probabilistic graphical models [9, 10, 51, 26], heuristics are unavoidable. These heuristics usually take the form of a search method and an evaluation measure. The evaluation measure estimates the quality of a given decomposition (a given conditional independence graph) and the search method determines which decompositions (which conditional independence graphs) are inspected. Often the search is guided by the value of the evaluation measure, since it is usually the goal to maximize (or to minimize) its value.

[22] develops a rigid foundation of a learning algorithm for possibilistic networks.

It starts from a comparison of the *nonspecificity* of a given multivariate possibility distribution to the distribution represented by a possibilistic network, thus measuring the loss of specificity, if the multivariate possibility distribution is represented by the network. The measure of nonspecificity can be derived from *Hartley information* [25], in contrast to some evaluation measures for learning probabilistic networks, which are based on *Shannon information* [45]. In order to arrive at an efficient algorithm, an approximation for this loss of specificity is derived, which can be computed locally on the maximal cliques of the network. As the search method a generalization of the optimum weight spanning tree algorithm is used. Several other heuristic local evaluation measures for learning possibilistic networks, which can be used with different search methods, are discussed in [3, 4]. Implementations based on these theoretical results have successfully been applied to the Danish Jersey cattle example. For details, see [18, 4].

6   Conclusions

In this paper we reviewed the state of the art of possibilistic graphical models w.r.t. evidence propagation and learning and indicated similarities and differences to probabilistic graphical models. To summarize, probabilistic approaches serve for the exact modeling of uncertain, but *precise* data, since imprecise data cannot be represented by a single probability distribution. Possibilistic approaches serve for the *approximate (information-compressed)* modeling of uncertain and/or *imprecise* data. Therefore, probabilistic and possibilistic graphical models are useful in quite different domains of knowledge representation, which makes them cooperative rather than competitive. A topic of future work is to study in which way probabilistic and possibilistic data, obtained from expert knowledge and/or databases of sample cases, can be combined and then be represented as the quantitative part of a unified type of graphical model.

7   Acknowledgments

The concepts and methods of possibilistic graphical modeling presented in this paper were applied within the CEC-ESPRIT III BRA 6156 DRUMS II project (Defeasible Reasoning and Uncertainty Management Systems) and in a cooperation with Deutsche Aerospace for the design of a data fusion tool [2]. Furthermore, the learning algorithms have been implemented during the design of a data mining tool that is developed in the research center of DaimlerChrysler in Ulm, Germany.

References

[1] S.K. Andersen, K.G. Olesen, F.V. Jensen, and F. Jensen. HUGIN — A Shell for Building Bayesian Belief Universes for Expert Systems. *Proc. 11th Int. J.*

Conf. on Artificial Intelligence (IJCAI'89, Detroit, MI, USA), 1080–1085. Morgan Kaufman, San Mateo, CA, USA 1989

[2] J. Beckmann, J. Gebhardt, and R. Kruse. Possibilistic Inference and Data Fusion. *Proc. 2nd European Congress on Fuzzy and Intelligent Technologies (EUFIT'94, Aachen, Germany)*, 46–47. Verlag Mainz, Aachen, Germany 1994

[3] C. Borgelt and R. Kruse. Evaluation Measures for Learning Probabilistic and Possibilistic Networks. *Proc. 6th IEEE Int. Conf. on Fuzzy Systems (FUZZ-IEEE'97, Barcelona, Spain)*, Vol. 2:1034–1038. IEEE Press, Piscataway, NJ, USA 1997

[4] C. Borgelt and R. Kruse. Some Experimental Results on Learning Probabilistic and Possibilistic Networks with Different Evaluation Measures. *Proc. 1st Int. J. Conf. on Qualitative and Quantitative Practical Reasoning (ECSQARU/FAPR'97, Bad Honnef, Germany)*, 71–85. Springer, Berlin, Germany 1997

[5] C. Borgelt and R. Kruse. Efficient Maximum Projection of Database-Induced Multivariate Possibility Distributions. *Proc. 7th IEEE Int. Conf. on Fuzzy Systems (FUZZ-IEEE'98, Anchorage, Alaska, USA)*, IEEE Press, Piscataway, NJ, USA 1997

[6] W. Buntine. Operations for Learning Graphical Models. *J. of Artificial Intelligence Research* 2:159–224, 1994

[7] L.M. de Campos, J. Gebhardt, and R. Kruse. Axiomatic Treatment of Possibilistic Independence. In: C. Froidevaux and J. Kohlas, eds. *Symbolic and Quantitative Approaches to Reasoning and Uncertainty (LNCS 946)*, 77–88. Springer, Berlin, Germany 1995

[8] E. Castillo, J.M. Gutierrez, and A.S. Hadi. *Expert Systems and Probabilistic Network Models*. Springer, New York, NY, USA 1997

[9] C.K. Chow and C.N. Liu. Approximating Discrete Probability Distributions with Dependence Trees. *IEEE Trans. on Information Theory* 14(3):462–467. IEEE Press, Piscataway, NJ, USA 1968

[10] G.F. Cooper and E. Herskovits. A Bayesian Method for the Induction of Probabilistic Networks from Data. *Machine Learning* 9:309–347. Kluwer, Dordrecht, Netherlands 1992

[11] R. Dechter. Bucket Elimination: A Unifying Framework for Probabilistic Inference. *Proc. 12th Conf. on Uncertainty in Artificial Intelligence (UAI'96, Portland, OR, USA)*, 211–219. Morgan Kaufman, San Mateo, CA, USA 1996

[12] A.P. Dempster. Upper and Lower Probabilities Induced by a Multivalued Mapping. *Ann. Math. Stat.* 38:325–339, 1967

[13] A.P. Dempster. Upper and Lower Probabilities Generated by a Random Closed Interval. *Ann. Math. Stat.* 39:957–966, 1968

[14] D. Dubois, S. Moral, and H. Prade. A Semantics for Possibility Theory based on Likelihoods. Annual report, CEC–ESPRIT III BRA 6156 DRUMS II, 1993

[15] D. Dubois and H. Prade. *Possibility Theory*. Plenum Press, New York, NY, USA 1988

[16] D. Dubois and H. Prade. Fuzzy Sets in Approximate Reasoning, Part 1: Inference

with Possibility Distributions. *Fuzzy Sets and Systems* 40:143–202. North Holland, Amsterdam, Netherlands 1991

[17] D. Dubois, H. Prade, and R.R. Yager, eds. *Readings in Fuzzy Sets for Intelligent Systems*. Morgan Kaufman, San Mateo, CA, USA 1993

[18] J. Gebhardt. *Learning from Data: Possibilistic Graphical Models*. Habilitation Thesis, University of Braunschweig, Germany 1997

[19] J. Gebhardt and R. Kruse. A New Approach to Semantic Aspects of Possibilistic Reasoning. In: M. Clarke, S. Moral, and R. Kruse, eds. *Symbolic and Quantitative Approaches to Reasoning and Uncertainty (LNCS 747)*, 151–159. Springer, Berlin, Germany 1993

[20] J. Gebhardt and R. Kruse. On an Information Compression View of Possibility Theory. *Proc. 3rd IEEE Int. Conf. on Fuzzy Systems (FUZZ-IEEE'94, Orlando, FL, USA)*, 1285–1288. IEEE Press, Picataway, NJ, USA 1994

[21] J. Gebhardt and R. Kruse. POSSINFER — A Software Tool for Possibilistic Inference. In: D. Dubois, H. Prade, and R. Yager, eds. *Fuzzy Set Methods in Information Engineering: A Guided Tour of Applications*, 407–418. J. Wiley & Sons, New York, NY, USA 1996

[22] J. Gebhardt and R. Kruse. Tightest Hypertree Decompositions of Multivariate Possibility Distributions. *Proc. Int. Conf. on Information Processing and Management of Uncertainty in Knowledge–Based Systems (IPMU'96)*, 923–927. Granada, Spain 1996

[23] J. Gebhardt and R. Kruse. Automated Construction of Possibilistic Networks from Data. *J. of Applied Mathematics and Computer Science*, 6(3):101–136, 1996

[24] D. Geiger, T.S. Verma, and J. Pearl. Identifying Independence in Bayesian Networks. *Networks* 20:507–534. J. Wiley & Sons, Chichester, England, 1990

[25] R.V.L. Hartley. Transmission of Information. *The Bell Systems Technical Journal* 7:535–563, 1928

[26] D. Heckerman, D. Geiger, and D.M. Chickering. Learning Bayesian Networks: The Combination of Knowledge and Statistical Data. *Machine Learning* 20:197–243, Kluwer, Dordrecht, Netherlands, 1995

[27] K. Hestir, H.T. Nguyen, and G.S. Rogers. A Random Set Formalism for Evidential Reasoning. In: I.R. Goodman, M.M. Gupta, H.T. Nguyen, and G.S. Rogers, eds. *Conditional Logic in Expert Systems*, 209–344. North-Holland, Amsterdam, Netherlands 1991

[28] E. Hisdal. Conditional Possibilities, Independence, and Noninteraction. *Fuzzy Sets and Systems* 1:283–297. North Holland, Amsterdam, Netherlands 1978

[29] V. Isham. An Introduction to Spatial Point Processes and Markov Random Fields. *Int. Statistics Review* 49:21–43, 1981

[30] F. Klawonn, J. Gebhardt, and R. Kruse. Fuzzy Control on the Basis of Equality Relations with an Example from Idle Speed Control. *IEEE Transactions on Fuzzy Systems* 3:336–350. IEEE Press, Piscataway, NJ, USA 1995

[31] R. Kruse, J. Gebhardt, and F. Klawonn. *Foundations of Fuzzy Systems*. J. Wiley

& Sons, Chichester, England 1994

[32] R. Kruse, E. Schwecke, and J. Heinsohn. *Uncertainty and Vagueness in Knowledge Based Systems: Numerical Methods.* Springer, Berlin, Germany 1991

[33] S.L. Lauritzen and D.J. Spiegelhalter. Local Computations with Probabilities on Graphical Structures and Their Application to Expert Systems. *Journal of the Royal Statistical Society, Series B* 2(50):157–224. Blackwell, Oxford, United Kingdom 1988

[34] H.T. Nguyen. Using Random Sets. *Information Science* 34:265–274, 1984

[35] J. Pearl. *Probabilistic Reasoning in Intelligent Systems: Networks of Plausible Inference (2nd edition).* Morgan Kaufmann, San Mateo, CA, USA 1992

[36] J. Pearl and A. Paz. Graphoids — A Graph Based Logic for Reasoning about Relevance Relations. In: B.D. Boulay et al., eds. *Advances in Artificial Intelligence 2*, 357–363. North-Holland, Amsterdam, Netherlands 1991

[37] L.K. Rasmussen. Blood Group Determination of Danish Jersey Cattle in the F-blood Group System. *Dina Research Report 8*, Dina Foulum, Tjele, Denmark 1992

[38] E.H. Ruspini. The Semantics of Vague Knowledge. *Rev. Internat. Systemique* 3:387–420, 1989

[39] E.H. Ruspini. Similarity Based Models for Possibilistic Logics. *Proc. 3rd Int. Conf. on Information Processing and Management of Uncertainty in Knowledge Based Systems (IPMU'96)*, 56–58. Granada, Spain 1990

[40] E.H. Ruspini. On the Semantics of Fuzzy Logic. *Int. J. of Approximate Reasoning* 5. North-Holland, Amsterdam, Netherlands 1991

[41] A. Saffiotti and E. Umkehrer. PULCINELLA: A General Tool for Propagating Uncertainty in Valuation Networks. *Proc. 7th Conf. on Uncertainty in Artificial Intelligence (UAI'91, Los Angeles, CA, USA)*, 323–331. Morgan Kaufman, San Mateo, CA, USA 1991

[42] G. Shafer. *A Mathematical Theory of Evidence.* Princeton University Press, Princeton, NJ, USA 1976

[43] G. Shafer and J. Pearl. *Readings in Uncertain Reasoning.* Morgan Kaufman, San Mateo, CA, USA 1990

[44] G. Shafer and P.P. Shenoy. *Local Computations in Hypertrees (Working Paper 201).* School of Business, University of Kansas, Lawrence, KS, USA 1988

[45] C.E. Shannon. The Mathematical Theory of Communication. *The Bell Systems Technical Journal* 27:379–423. 1948

[46] P.P. Shenoy. A Valuation-based Language for Expert Systems. *Int. J. of Approximate Reasoning* 3:383–411. North-Holland, Amsterdam, Netherlands 1989

[47] P.P. Shenoy. Valuation Networks and Conditional Independence. *Proc. 9th Conf. on Uncertainty in AI (UAI'93)*, 191–199. Morgan Kaufman, San Mateo, CA, USA 1993

[48] P.P. Shenoy and G.R. Shafer. Axioms for Probability and Belief-Function Propagation. In: R.D. Shachter, T.S. Levitt, L.N. Kanal, and J.F. Lemmer. *Uncertainty*

in Artificial Intelligence 4, 169–198. North Holland, Amsterdam, Netherlands 1990

[49] P. Smets and R. Kennes. The Transferable Belief Model. *Artificial Intelligence* 66:191–234. Elsevier, Amsterdam, Netherlands 1994

[50] D. Spiegelhalter, A. Dawid, S. Lauritzen, and R. Cowell. Bayesian Analysis in Expert Systems. *Statistical Science* 8(3):219–283, 1993

[51] T.S. Verma and J. Pearl. An Algorithm for Deciding if a Set of Observed Independencies has a Causal Explanation. *Proc. 8th Conf. on Uncertainty in Artificial Intelligence (UAI'92, Stanford, CA, USA)*, 323–330. Morgan Kaufman, San Mateo, CA, USA 1992

[52] P. Walley. *Statistical Reasoning with Imprecise Probabilities*. Chapman & Hall, New York, NY, USA 1991

[53] P.Z. Wang. From the Fuzzy Statistics to the Falling Random Subsets. In: P.P. Wang, ed. *Advances in Fuzzy Sets, Possibility and Applications*, 81–96. Plenum Press, New York, NY, USA 1983

[54] J. Whittaker. *Graphical Models in Applied Multivariate Statistics*. J. Wiley & Sons, Chichester, England 1990

[55] L.A. Zadeh. The Concept of a Linguistic Variable and Its Application to Approximate Reasoning. *Information Sciences* 9:43–80, 1975

[56] L.A. Zadeh. Fuzzy Sets as a Basis for a Theory of Possibility. *Fuzzy Sets and Systems* 1:3–28. North-Holland, Amsterdam, Netherlands 1978

[57] N.L. Zhang and D. Poole. Exploiting Causal Independence in Bayesian Network Inference. *Journal of Artificial Intelligence Research* 5:301–328, 1996

AN OVERVIEW OF POSSIBILISTIC LOGIC AND ITS APPLICATION TO NONMONOTONIC REASONING AND DATA FUSION

S. Benferhat, D. Dubois and H. Prade
Paul Sabatier University, Toulouse, France

Abstract: This paper provides a brief survey of possibilistic logic as a simple and efficient tool for handling nonmonotonic reasoning and data fusion. In nonmonotonic reasoning, Lehmann's preferential System P is known to provide reasonable but very cautious conclusions, and in particular, preferential inference is blocked by the presence of "irrelevant" properties. When using Lehmann's rational closure, the inference machinery, which is then more productive, may still remain too cautious. These two types of inference can be represented using a possibility theory-based semantics. The paper proposes several safe ways to overcome the cautiousness of these systems. One of these ways takes advantage of (contextual) independence assumptions of the form: the fact that δ is true (or is false) does not affect the validity of the rule "normally if α then β". The modelling of such independence assumptions is discussed in the possibilistic framework. This paper presents a general approach for fusing several ordered belief bases provided by different sources according to various modes. More precisely, the paper provides the syntactic counterparts of different ways of aggregating possibility distributions, well-known at the semantic level.

1. Introduction

Possibilistic logic (Dubois, Lang and Prade, 1994) provides a simple and efficient tool for performing uncertainty reasoning, where uncertain information are encoded under the form of a set of possibilistic logic formulas. Possibilistic knowledge bases are made of classical logic formulas weighted by lower bounds of possibility or necessity measures. When a possibilistic knowledge base is consistent in the classical sense (i.e., without considering the weights of the formulas), possibilistic inference is monotonic. Moreover, the level of certainty of a conclusion is equal to the greatest value attached to an inference path leading to this conclusion, the value of an inference path being the weight associated with the least certain formula(s) used in the path. When the knowledge base is inconsistent, a nonmonotonic behavior may take place. Indeed, a level of partial inconsistency is computed for the knowledge base as the greatest level of certainty attached to the derivation of the contradiction from the knowledge base. Conclusions obtained with levels of certainty strictly higher than the level of partial inconsistency are still valid since they are produced from a consistent and reliable subpart of the knowledge base. Thus, when new weighted formulas are added to the knowledge base, its level of inconsistency may become higher, and some previous conclusions with levels of certainty smaller than the new level of inconsistency may be invalidated, while new, possibly opposite, conclusions can be obtained from the new knowledge base.

In previous papers (Benferhat et al., 1992; 1997a), the authors have shown that a simple semantics can be provided for nonmonotonic inferences in the framework of possibility theory. It has been proposed to view each conditional assertion $\alpha \to \beta$ as a constraint expressing that the situation where α and β is true has a greater plausibility than the one where α and $\neg\beta$ is true. In other words, in the context where α is true, β is more possible or plausible than $\neg\beta$, i.e., the exceptional situation $\alpha \wedge \neg\beta$ is strictly less possible than the normal state of affairs which is $\alpha \wedge \beta$. Possibilisitic formulas can be hence viewed as expressing a rule with possible exceptions, the higher the certainty level of the formula (expressed in terms of necessity measures), the more exceptional are the possible exceptions to the rule, or in other words, the more abnormal it is to encounter exceptions. This paper expands the contents of two previous articles (Benferhat et al., 1996a, b). It proposes several safe ways to remedy some drawbacks of the well known System P and of the rational closure method (Kraus et al., 1990). One of these ways takes advantage of (contextual) independence assumptions of the form: the fact that δ is true (or is false) does not affect the validity of the rule "normally if α then β". We investigate how to model such independence assumptions in the possibilistic framework.

Another issue considered in this paper is that of databases merging (Baral et al., 1992; Cholvy, 1998; Lin and Mendelzon, 1992). At the semantic level, fusing two plausibility relations over a set of interpretations, can be achieved if the relations are commensurate that is, represented by mappings from interpretations to the same plausibility scale. In possibility theory, many fusion modes can be defined, especially disjunctive and conjunctive modes, based on fuzzy set theoretic operations. Syntactic fusion modes should be in accordance with semantic ones in order to be meaningful.

This paper is organised as follows: the next section gives basic definitions of possibility theory. Section 3 shows that both hard rules (rules which do not admit exceptions) and default rules (rules having exceptions) can be encoded in possibility theory. Several inference relations are proposed in this section, and two of them extend the preferential inference in a safe way. Possibilistic independence and its application to default reasoning are discussed in Section 4. The last section is devoted to semantically meaningful syntactic fusion and revision methods.

2. Possibility theory

In the following, $\mathcal{L}$ denotes a finite propositional language. Propositional variables are denoted by lower case Roman letters a, b, c... *Formulae* of $\mathcal{L}$ are denoted by Greek letters α, β, δ,... An *interpretation* (called also a *world*) for $\mathcal{L}$ is an assignment of a truth value in {T, F} to each formula of $\mathcal{L}$ in accordance with the classical rules of propositional calculus; we denote by Ω the set of all such interpretations. An interpretation ω is a *model* of a formula α, and we write $\omega \models \alpha$ iff $\omega(\alpha) = T$, where T represents formulas satisfied by each interpretation (*tautology*), and $\perp$ denotes any inconsistent formula. As usual a formula α is said to be consistent if and only if it has at least one model, and is said to be inconsistent otherwise. We denote by $[\alpha]$ the set of models of α.

2.1. Possibility distributions

We give here some elementary definitions of possibility theory (Zadeh, 1978), (Dubois and Prade, 1988). The basic object of possibility theory is the *possibility distribution*, which is a mapping from the set of classical interpretations Ω to the interval [0,1]. A possibility distribution corresponds to a ranking on Ω, such that the most plausible worlds get the highest value. The possibility distribution π represents the available knowledge about where the real world is. By convention, $\pi(\omega) = 1$ means that it is totally possible for ω to be the real world, $\pi(\omega) > 0$ means that ω is only somewhat possible, while $\pi(\omega) = 0$ means that ω is certainly not the real world. The possibility distribution π is said to be *normal* if there exists at least one interpretation ω which is totally possible, namely $\pi(\omega) = 1$. However, in general there may exist several distinct interpretations which are totally possible. This *normalisation condition* reflects the consistency of the available knowledge represented by this possibility distribution. The inequality $\pi(\omega) > \pi(\omega')$ means that the situation ω is a priori more plausible than ω'. Note that if we choose a threshold a and consider $\{\omega \mid \pi(\omega) > a\}$ we get what Lewis (1973) calls a "sphere" around the most plausible states of the world. Hence π encodes a system of spheres, a unique one for the whole set of interpretations. A possibility distribution π induces two mappings grading respectively the possibility and the certainty of a formula α:

– the possibility degree $\Pi(\alpha) = \sup\{\pi(\omega) \mid \omega \models \alpha\}$ which evaluates to what extent α is consistent with the available knowledge expressed by π. Note that we have:
$$\forall \alpha \; \forall \beta \;\; \Pi(\alpha \vee \beta) = \max(\Pi(\alpha), \Pi(\beta));$$
– the necessity (or certainty) degree $N(\alpha) = \inf\{1 - \pi(\omega) \mid \omega \models \neg\alpha\}$ which evaluates to what extent α is entailed by the available knowledge. We have:
$$\forall \alpha \; \forall \beta \;\; N(\alpha \wedge \beta) = \min(N(\alpha), N(\beta)).$$

Certainty and possibility measures are related by the usual duality equation $N(\alpha) = 1 - \Pi(\neg\alpha)$ between what is possible and what is necessarily true. Moreover, note that, contrarily to probability theory $N(\alpha)$ and $N(\neg\alpha)$ (resp. $\Pi(\alpha)$ and $\Pi(\neg\alpha)$) are not functionally related: we only have (for normal possibility distributions) $\min(N(\alpha), N(\neg\alpha)) = 0$ (resp. $\max(\Pi(\alpha), \Pi(\neg\alpha)) = 1$). Conversely, a sub-normal possibility distribution would lead to both $N(\alpha) > 0$ and $N(\neg\alpha) > 0$, expressing a tentative acceptance of α and $\neg\alpha$ simultaneously, which is an inconsistent situation. It leaves room for representing complete ignorance in an unbiased way: α is ignored whenever $\Pi(\alpha) = \Pi(\neg\alpha) = 1$, i.e., neither α nor $\neg\alpha$ is somewhat abnormal. Moreover, given a normal possibility distribution π, the set of propositions α such that $N(\alpha) > 0$ is deductively closed and forms what Gärdenfors (1988) calls a belief set. It is the set of (more or less) accepted propositions in the epistemic state described by π. $N(\alpha) = 1$ expresses absolute certainty that p is true. Such measures of certainty as N appear under various guises in the pioneering works of Shackle (1961), L.J. Cohen (1977), Rescher (1976) and more recently in Spohn (1988).

A distinctive characteristic of the functions N and Π lies in their ordinal nature, i.e., the unit interval is only used to rank-order the various possible situations in Ω, in terms of their compatibility with the knowledge as encoded by the possibility distribution π. Indeed, a qualitative necessity (resp. possibility) relation can also be defined for any pair of formulas α and β as $\alpha \geq_N \beta \Leftrightarrow N(\alpha) \geq N(\beta)$ (resp. $\alpha \geq_\Pi \beta \Leftrightarrow \Pi(\alpha) \geq \Pi(\beta)$). Note that $\geq_N$ and $\geq_\Pi$ are complete pre-orders verifying moreover $\top >_N \bot$ (resp. $\top >_\Pi \bot$). Another important notion in possibility theory is the *principle of minimum specificity*. A possibility distribution π is said to be *more specific* (Yager, 1992) than another π' if and only if for each interpretation ω we have $\pi(\omega) \leq \pi'(\omega)$ and there exists at least one interpretation ω' such that $\pi(\omega') < \pi'(\omega')$. In other words, π is more informative than π'.

Given a set of constraints restricting a feasible subset of possibility distributions, the best representative is the least specific feasible possibility distribution, which assigns the highest degree of possibility to each world, since it is the least committed one.

As for probability theory, a notion of conditioning can be defined for possibility and necessity measures, by means of an equation similar to Bayesian conditioning (Hisdal, 1978) (Dubois and Prade, 1990)

$$\Pi(\alpha \wedge \beta) = \min(\Pi(\beta \mid \alpha), \Pi(\alpha))$$

when $\Pi(\alpha) > 0$. $\Pi(\beta \mid \alpha)$ is defined as the greatest solution to the previous equation in accordance with the minimum specificity principle. It leads to

$$\Pi(\beta \mid \alpha) = 1 \text{ if } \Pi(\alpha \wedge \beta) = \Pi(\beta) \qquad (\text{i.e., } \Pi(\alpha \wedge \beta) \geq \Pi(\alpha \wedge \neg\beta))$$
$$= \Pi(\alpha \wedge \beta) \qquad \text{otherwise} \qquad (\text{i.e., } \Pi(\alpha \wedge \beta) < \Pi(\alpha \wedge \neg\beta))$$

when $\Pi(\alpha) > 0$. If $\Pi(\alpha) = 0$, then by convention $\Pi(\beta \mid \alpha) = 1$, $\forall \beta \neq \perp$. The conditional necessity measure is simply defined as $N(\beta \mid \alpha) = 1 - \Pi(\neg\beta \mid \alpha)$. Thus $N(\beta \mid \alpha) > 0$ means that in the context α, β is accepted. It can be easily checked that $N(\beta \mid \alpha) > 0$ iff $\Pi(\alpha \wedge \beta) > \Pi(\alpha \wedge \neg\beta)$, which means that accepting β in the context α, is equivalent to saying that the situation where $\alpha \wedge \beta$ is true is more plausible then the situation where $\alpha \wedge \neg\beta$ is true. Finally, a possibilistic entailment $\models_\pi$ can be defined at a semantical level in the spirit of Shoham (1988)'s proposal: $\alpha \models_\pi \beta \Leftrightarrow$ all the worlds which maximize π, among those which satisfy α, satisfy β.

We restrict this definition to propositions α such that $\Pi(\alpha) > 0$. In the following, we denote by $[\alpha]_\pi$ the set of interpretations $A \subset \Omega$ such that for each $\omega \in A$ we have $\pi(\omega) = \Pi(\alpha)$. It has been shown that the inference relation $\models_\pi$ is nonmonotonic, and it can be established that (Dubois and Prade, 1991)

$$\alpha \models_\pi \beta \qquad \text{if and only if } \{\omega \models \alpha \mid \pi(\omega) = \Pi(\alpha) > 0\} \subseteq \{\omega \models \beta\}$$
$$\text{if and only if } N(\beta \mid \alpha) > 0$$
$$\text{if and only if } \Pi(\alpha \wedge \beta) > \Pi(\alpha \wedge \neg\beta).$$

2.2. A possibilistic belief base and its induced possibility distribution

A possibilistic logic formula is a pair (α, a) where α is a propositional logic formula and a an element of the semi-open real interval $(0,1]$, which estimates to what extent it is certain that α is true considering the available, possibly incomplete information about the world. More formally (α, a) is a syntactic way of encoding the semantic constraint $N(\alpha) \geq a$.

The semantics of the weighted formula (α, a) is represented by the fuzzy set of models $M(\alpha, a)$ defined by

$$\mu_{M(\alpha,a)}(\omega) = 1 \text{ if } \omega \in [\alpha];$$
$$\mu_{M(\alpha,a)}(\omega) = 1 - a \text{ if } \omega \notin [\alpha]. \qquad (1)$$

where $[\alpha]$ denotes the set of models of α (in the classical sense). In other words, the interpretations compatible with (α, a) are restricted by the above fuzzy set $M(\alpha, a)$ whose membership function is interpreted as a possibility distribution. The interpretations in $[\alpha]$, where α is true, are considered as fully possible while the interpretations outside $[\alpha]$ (where α is false) are all the more possible as a is smaller, i.e., the piece of knowledge is less certain. Any possibility distribution π satisfying the constraint $N(\alpha) \geq a$ is such that

$$\forall \omega, \ \pi(\omega) \leq \mu_{M(\alpha,a)}(\omega)$$

since $\Pi(\neg\alpha) \leq 1 - a \Leftrightarrow N(\alpha) \geq a$. Hence (1) is justified by the minimal specificity principle since it yields the least specific possibility distribution $\pi = \mu_{M(\alpha,a)}$, i.e., the one with the greatest possibility degrees, compatible with (α,a).

In case of several pieces of knowledge (α_i,a_i), for i=1,n, forming a knowledge base Σ, we combine them in accordance with the minimal specificity principle. Consequently, a possibility distribution π_Σ is attached to the set of possibilistic formulas Σ, and built by applying the minimum operator to the membership functions $\mu_{M(\alpha_i,a_i)}$, for i =1,n, namely

$$\pi_\Sigma(\omega) = \min_{i=1,n} \mu_{M(\alpha_i,a_i)}(\omega).$$

It can be checked that the necessity measure N_Σ induced from π_Σ by

$$N_\Sigma(\beta) = 1 - \max\{\pi_\Sigma(\omega), \omega \models \neg\beta\},$$

is the smallest necessity measure satisfying the constraints $N(\alpha_i) \geq a_i$ for i=1,n. Indeed, since π_Σ is the least specific possibility distribution satisfying the constraints $N(\alpha_i) \geq a_i$ for i=1,n, then for any π which satisfies the same constraints we have: $\forall\omega, \pi_\Sigma(\omega) \geq \pi(\omega)$. This implies that $N_\Sigma(\beta)=1 - \max\{\pi_\Sigma(\omega), \omega \models \neg\beta\} \leq N_\pi(\beta)=1 - \max\{\pi(\omega), \omega \models \neg\beta\}$. Thus propositions are not made more certain than they are. This is nothing but another formulation of the minimal specificity principle in terms of certainty. Generally, a strict inequality $N_\Sigma(\alpha_i) > a_i$ may be obtained for some i. Indeed, for instance, the set of constraints $\{N(\alpha) \geq a, N(\neg\alpha\vee\beta)\geq a, N(\beta) \geq b\}$ with $a > b$, entails $N_\Sigma(\beta)\geq a>b$. If π_Σ is not normal, $\exists\gamma\neq\perp,T$ such that $N(\gamma)>0$ and $N(\neg\gamma)>0$, which expresses the inconsistency of Σ.

Example 1:

Let us take an illustrative example:

$$\Sigma = \{(\neg\alpha \vee \beta, 1), (\neg\alpha \vee \delta, 0.7), (\neg\beta \vee \delta, 0.4), (\alpha, 0.5), (\beta, 0.8)\}$$

It induces the constraints,

$$\pi(\omega) \leq \mu_{M(\neg\alpha \vee \beta, 1)}(\omega) \qquad \forall\omega \models \alpha \wedge \neg\beta, \pi(\omega) = 0$$
$$\pi(\omega) \leq \mu_{M(\neg\alpha \vee \delta, 0.7)}(\omega) \qquad \forall\omega \models \alpha\wedge\neg\delta, \pi(\omega) \leq 0.3$$
$$\pi(\omega) \leq \mu_{M(\neg\beta \vee \delta, 0.4)}(\omega) \quad\Leftrightarrow\quad \forall\omega \models \beta\wedge\neg\delta, \pi(\omega) \leq 0.6$$
$$\pi(\omega) \leq \mu_{M(\alpha, 0.5)}(\omega) \qquad \forall\omega \models \neg\alpha, \pi(\omega) \leq 0.5$$
$$\pi(\omega) \leq \mu_{M(\beta, 0.8)}(\omega) \qquad \forall\omega \models \neg\beta, \pi(\omega) \leq 0.2$$

The corresponding possibility distribution π_Σ, is

$$\pi_\Sigma(\alpha\beta\delta) = 1; \pi_\Sigma(\alpha\beta\neg\delta) = 0.3; \pi_\Sigma(\alpha\neg\beta\delta) = \pi_\Sigma(\alpha\neg\beta\neg\delta) = 0;$$
$$\pi_\Sigma(\neg\alpha\beta\delta) = \pi_\Sigma(\neg\alpha\beta\neg\delta) = 0.5; \pi_\Sigma(\neg\alpha\neg\beta\delta) = \pi_\Sigma(\neg\alpha\neg\beta\neg\delta) = 0.2; \qquad ■$$

The semantic entailment of (α, a) from a possibilistic knowledge base Σ is defined by

$$\Sigma \models (\alpha, a) \Leftrightarrow \forall\omega, \pi_\Sigma(\omega) \leq \mu_{M(\alpha, a)}(\omega).$$

It can be checked that

$$\Sigma \models (\alpha, a) \qquad \Leftrightarrow \forall\omega \models \neg\alpha, \pi_\Sigma(\omega) \leq 1 - a$$
$$\Leftrightarrow \Pi_\Sigma(\neg\alpha) \leq 1 - a$$
$$\Leftrightarrow N_\Sigma(\alpha) \geq a.$$

In the following we may omit the index Σ, when writing Π and N.

The possibility distribution in the example above is normalized, i.e., $\exists \omega, \pi_\Sigma(\omega) = 1$. It means that Σ is fully consistent since there is at least one interpretation in agreement with Σ which is completely possible. More generally we define the degree of inconsistency of Σ by

$$\text{Inc}(\Sigma) = 1 - \max_\omega \pi_\Sigma(\omega).$$

It can be also established that $\text{Inc}(\Sigma) > 0 \Leftrightarrow \Sigma^*$ is inconsistent, where Σ^* is the classical knowledge base obtained from Σ by deleting the weights. In other words, the consistency of Σ^* ensures that π_Σ is normalised and conversely.

When Σ is partially inconsistent $(\text{Inc}(\Sigma) > 0)$, the semantic entailment can be refined by considering as legitimally entailed only those propositions to which a degree of certainty, higher than the level of inconsistency, can be attached. Namely:

$$\exists\, a > \text{Inc}(\Sigma),\ \Sigma \models_\pi (\alpha, a) \Leftrightarrow \forall \omega,\ \pi_\Sigma(\omega) \le \mu_{M(\alpha,\, a)}(\omega)$$
$$\Leftrightarrow \Pi(\neg\alpha) \le 1 - a < 1 - \text{Inc}(\Sigma) = \Pi(\alpha)$$
$$\Leftrightarrow \Pi(\neg\alpha) < \Pi(\alpha).$$

since $1 - \text{Inc}(\Sigma) = \max(\Pi(\alpha), \Pi(\neg\alpha))$. This type of entailment can be studied directly in terms of a plausibility ordering among distributions, encoded by a possibility distribution.

2.3 - Possibilistic resolution principle

In this section we suppose that weighted formulas are put under the form of weighted *clauses*; this can be done without loss of expressivity, due to the compositionality of necessity measures with respect to conjunction. The following resolution rule (Dubois and Prade, 1987)

$$(\alpha \vee \beta,\, a)\, ;\, (\neg\alpha \vee \delta,\, b) \vdash (\beta \vee \delta,\, \min(a,\, b))$$

is valid in possibilistic logic.

In order to compute the maximal certainty degree which can be attached to a formula according to the constraints expressed by a knowledge base Σ, for instance δ in the above example of Section 2.2., just add to Σ the clause(s) obtained by refuting the proposition to evaluate, with a necessity degree equal to 1; here add $(\neg\delta, 1)$. Then it can be shown that any lower bound obtained on $\bot$, by resolution, is a lower bound of the necessity of the proposition to evaluate. See (Dubois et al., 1987) for an ordered search method which guarantees that we obtain the greatest derivable lower bound on $\bot$. It can be shown (e.g., Dubois et al., 1994)), that this greatest derivable lower bound on $\bot$ is nothing but the inconsistency degree $\text{Inc}(\Sigma \cup \{(\neg\delta, 1)\})$ where δ is the proposition to establish. Denote $\vdash$ the syntactic inference in possibilistic logic, based on refutation and resolution. Then the equivalence $\Sigma \vdash (\alpha, a) \Leftrightarrow \Sigma \models (\alpha, a)$ holds, i.e., $\vdash$ is sound and complete for refutation with respect to the semantics recalled in the previous sections (Dubois et al., 1994).

Example 1 (Continued):

In Example 1, we have the following derivation

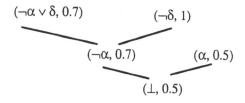

$$(\neg\alpha \vee \delta,\, 0.7) \qquad\qquad (\neg\delta,\, 1)$$
$$(\neg\alpha,\, 0.7) \qquad\qquad (\alpha,\, 0.5)$$
$$(\bot,\, 0.5)$$

i.e., $N(\delta) \geq 0.5$ and indeed it can be checked that using the possibility distribution $\pi = \pi_\Sigma$ of Section 3.1, we have $\Pi(\neg\delta) = \sup\{\pi(\omega), \omega \models \neg\delta\} = 0.5$, in fact $\Pi(\neg\delta) \leq 0.5$ since π is the greatest possibility distribution compatible with Σ, and then $N(\delta) = 1 - \Pi(\neg\delta) \geq 0.5$. In case of partial inconsistency of Σ, a refutation carried out in a situation where $\text{Inc}(\Sigma \cup \{(\neg\delta, 1)\}) = a > \text{Inc}(\Sigma)$ yields the non-trivial conclusion (δ, a), only using formulas whose degree of certainty is strictly greater than the level of inconsistency of the base (and particularly, at least equal to a). The possibilistic inference method is as efficient as classical logic refutation by resolution, and has been implemented in the form of an A*-like algorithm (Dubois et al., 1987). It has been shown (Lang, 1991) that the possibilistic entailment can be achieved with only $\text{Log}(n)$ satisfiability tests, where n is the number of uncertainty levels appearing in Σ.

3. Handling default and strict knowledge in possibility theory

3.1. Universal possibilistic consequence

By a conditional information (we call it also a conditional assertion or a default rule) we mean a generic rule of the form "generally, if α then β" having possibly some exceptions. These rules are denoted by "$\alpha \rightarrow \beta$" where $\rightarrow$ is a *non-classical* arrow relating two classical formulas. In the whole paper the arrow $\rightarrow$ has this non-classical meaning. A *default base* is a set $\Delta = \{\alpha_i \rightarrow \beta_i, i = 1, ..., n\}$ of default rules. The material implication is denoted by $\Rightarrow$, and is sometimes written as a disjunction. This material implication is used to encode strict rules (called also hard rules) of the form "if α_i is observed, then β_i is *always* true". We denote by $W = \{\alpha_i \Rightarrow \beta_i \ / \ i = 1,m\}$ a set of strict rules.

In (Benferhat et al., 1992), it has been proposed to view each conditional assertion $\alpha \rightarrow \beta$ as a constraint expressing that the situation where α and β is true has a greater plausibility than the one where α and $\neg\beta$ is true. In other words, we express that the exceptional situation $\alpha \wedge \neg\beta$ is strictly less possible than the normal state of affairs which is $\alpha \wedge \beta$, by the strict inequality

$$\Pi(\alpha \wedge \beta) > \Pi(\alpha \wedge \neg\beta).$$

All possibility measures satisfying this inequality do express that if α then β is normally true. They correspond to all epistemic states where the rule is accepted. This minimal requirement is very natural since it guarantees that all rules of the default base are acknowledged.

Moreover, hard rules of the form "all α's are β's" are modelled in possibility theory by the condition $\Pi(\alpha \wedge \neg\beta) = 0$ (Benferhat, 1994). Equivalently, any situation where $\alpha \wedge \neg\beta$ is true is impossible. Here, it exactly coincides with the classical logic treatment.

A body of knowledge $(\Delta = \{\alpha_i \rightarrow \beta_i, i = 1,n\}, W = \{\alpha_i \Rightarrow \beta_i, i = 1,m\})$ with consistent conditions (i.e., $\forall i, \alpha_i \neq \bot$), can thus be viewed as a family of constraints $\mathcal{C}(\Delta,W)$ restricting a family $\Pi(\Delta,W)$ of possibility distributions. Elements of $\Pi(\Delta,W)$ are called *compatible* with (Δ,W) and are defined as:

Definition 1: A possibility distribution π is said to be *compatible* with (Δ,W) iff the following conditions are satisfied:
    (i) for each hard rule $\alpha_i \Rightarrow \beta_i$ of W, we have: $\Pi(\alpha_i \wedge \neg\beta_i) = 0$,
    (ii) for each default rule $\alpha_i \rightarrow \beta_i$ of Δ, we have: $\Pi(\alpha_i \wedge \beta_i) > \Pi(\alpha_i \wedge \neg\beta_i)$,

where $\prod$ is defined from π.

Note that $\prod(\Delta,W)$ can be empty, and in this case our beliefs (Δ,W) are said to be inconsistent. A typical example of an inconsistent default base is $\Delta = \{\alpha \to \beta, \alpha \to \neg\beta\}$. In general, there are several possibility distributions which are compatible with (Δ,W). The question now is how to define which conditionals $\alpha \to \beta$ are entailed from our beliefs (Δ,W)? A first way of defining a nonmonotonic consequence relation consists in considering all the possibility distributions of $\prod(\Delta,W)$, namely:

Definition 2: A conditional assertion $\alpha \to \beta$ is said to be a *universal possibilistic consequence* of (Δ,W), denoted by $(\Delta,W) \models_{\forall\prod} \alpha \to \beta$, if and only if β is a possibilistic consequence of α in each possibility distribution of $\prod(\Delta,W)$, namely:

$$(\Delta,W) \models_{\forall\prod} \alpha \to \beta \qquad \text{iff} \qquad \forall \pi \in \prod(\Delta,W), \alpha \models_\pi \beta.$$

In Definition 2, the set of plausible conclusions are conditional assertions. In the following, we will use the notation $\alpha \models_{\forall\prod,(\Delta,W)} \beta$ (when there is no confusion on the body of background knowledge, we simply write $\alpha \models_{\forall\prod} \beta$) to denote that a propositional formula β is a plausible conclusion of the generic knowledge (Δ,W) and the observation α. $\alpha \models_{\forall\prod} \beta$ is equivalently defined by $(\Delta,W) \models_{\forall\prod} \alpha \to \beta$.

Example 2: Let us consider the following (usual) set of default rules "generally, birds fly", "generally, penguins do not fly", "all penguins are birds", symbolically written as $\Delta=\{b \to f, p \to \neg f\}$, and $W = \{p \Rightarrow b\}$. We are interested to know if a given bird "Tweety" which is a penguin flies or not. Let Ω be the following set of possible interpretations:

$$\Omega = \{ \; \omega_0: \neg b \wedge \neg f \wedge \neg p, \; \omega_1: \neg b \wedge \neg f \wedge p, \; \omega_2: \neg b \wedge f \wedge \neg p, \; \omega_3: \neg b \wedge f \wedge p,$$
$$\omega_4: b \wedge \neg f \wedge \neg p, \; \omega_5: b \wedge \neg f \wedge p, \; \omega_6: b \wedge f \wedge \neg p, \; \omega_7: b \wedge f \wedge p\}$$

Each possibility distribution π of $\prod(\Delta, W)$ must satisfies the following constraints:

$$\prod(b \wedge f) > \prod(b \wedge \neg f)$$
$$\prod(p \wedge \neg f) > \prod(p \wedge f)$$
$$\prod(p \wedge \neg b) = 0.$$

These constraints induce the following set of constraints C' on models:

$$C'_1: \max(\pi(\omega_6), \pi(\omega_7)) > \max(\pi(\omega_4), \pi(\omega_5))$$
$$C'_2: \max(\pi(\omega_5), \pi(\omega_1)) > \max(\pi(\omega_3), \pi(\omega_7))$$
$$C'_3: \pi(\omega_1) = \pi(\omega_3) = 0.$$

Using C'_3, the constraint C'_2 is simplified in:

$$C'_2: \pi(\omega_5) > \pi(\omega_7).$$

Now assume that $p \wedge b \not\models_{\forall\prod} \neg f$. It means that there exists at least one possibility distribution π compatible with (Δ, W) such that:

$$\prod(p \wedge b \wedge f) \geq \prod(p \wedge b \wedge \neg f).$$

Or equivalently: $\qquad C'_4: \pi(\omega_7) \geq \pi(\omega_5).$

This is not possible since it directly contradicts the constraint C'_2. Hence, the inference:

$p \wedge b \models_{\forall\prod} \neg f$ is valid. ∎

The universal consequence, when it produces conditional assertions, is always monotonic. Namely if $(\Delta, W) \models_{\forall\Pi} b \rightarrow w$ then $(\Delta', W) \models_{\forall\Pi} b \rightarrow w$ is still valid for any Δ' containing Δ. However, the universal possibilistic consequence relation defined between propositional formulas is nonmonotonic. Indeed, we can easily check from the Example 2 that birds fly (i.e., $b \models_{\forall\Pi} f$), but birds which are penguins do not fly (i.e., $p \wedge b \models_{\forall\Pi} \neg f$ and $p \wedge b \not\models_{\forall\Pi} f$). Several authors have tried to delimit natural properties of a nonmonotonic consequence relation, often denoted by $\vdash\!\!\sim$, likely to achieve a satisfactory treatment of plausible reasoning in the presence of incomplete information. Typical of this kind of research are the works of Gabbay (1985), Makinson (1989), Gärdenfors and Makinson (1994) and Kraus, Lehmann and Magidor (1990). The latter have proposed a set of axioms called System P (P as "preferential"). The inference induced by System P, denoted by $\vdash\!\!\sim_P$, corresponds to the ε-semantics of Adams (1975) based on the probabilistic (infinitesimal) interpretation of default rules (Pearl, 1990). It turns out that $\models_{\forall\Pi}$ is equivalent to $\vdash\!\!\sim_P$ (Dubois and Prade, 1995):

Proposition 1:   If $W = \emptyset$ then $\alpha \models_{\forall\Pi} \beta$ if and only if $\alpha \vdash\!\!\sim_P \beta$.

3.2. Irrelevance Problem

The universal possibilistic consequence relation, even if it produces acceptable and safe conclusions, is very cautious and suffers, as System P, from a so-called *"irrelevance"* problem.[1] This problem can be described in the following way: if a formula δ is a plausible consequence of α, and if a formula β has "nothing to do" (namely is irrelevant to) with α or δ, then δ cannot be deduced from $\alpha \wedge \beta$, while such a deduction is expected. The word "irrelevant" is not well-defined in the literature and in Section 4 we try to formalize this notion in details. In this subsection we will only consider a particular case of irrelevance which corresponds to the case where β is a formula composed of propositional symbols which do not appear in the default base. This situation is illustrated by the following example:

Example 3: Let us only consider one default rule $\Delta = \{b \rightarrow f\}$. Intuitively, from this default base, given a red bird we like to deduce that it flies too, even if the knowledge base does not say explicitly that such birds fly. The representation used here is parsimonious since it does not give a complete description of the world, thus some conclusions must be implicitly deduced. The universal possibilistic consequence relation unfortunately cannot infer that red birds fly. Indeed, assume that our language only contains three propositional symbols $\{b,f,r\}$ where r is for red, then:

$$\Omega = \{ \ \omega_0: \neg b \wedge \neg f \wedge \neg r, \ \omega_1: \neg b \wedge \neg f \wedge r, \ \omega_2: \neg b \wedge f \wedge \neg r, \ \omega_3: \neg b \wedge f \wedge r,$$
$$\omega_4: b \wedge \neg f \wedge \neg r, \ \omega_5: b \wedge \neg f \wedge r, \ \omega_6: b \wedge f \wedge \neg r, \ \omega_7: b \wedge f \wedge r\}$$

Each π of $\Pi(\Delta, W)$ satisfies the constraint:

$$C_1: \Pi(b \wedge f) > \Pi(b \wedge \neg f)$$

---

[1] The name "irrelevance" is partially a misnomer, but commonly used.

or equivalently (on the models):

$$C'_1: \max(\pi(\omega_6), \pi(\omega_7)) > \max(\pi(\omega_4), \pi(\omega_5)).$$

We can easily check that the three following possibility distributions:

- $\pi_1(\omega_6) = \pi_1(\omega_7) = 1, \quad \pi_1(\text{otherwise}) = \alpha < 1$
- $\pi_2(\omega_6) = 1; \quad \pi_2(\omega_5) = \alpha < 1, \quad\quad\quad \pi_2(\text{otherwise}) = \beta < \alpha$
- $\pi_3(\omega_6) = 1 \quad \pi_3(\omega_7) = \pi_3(\omega_5) = \alpha < 1, \pi_3(\text{otherwise}) = \beta < \alpha$

are all compatible with (Δ, W), and we can easily verify that:

$$b \wedge r \vDash_{\pi_1} f, \, b \wedge r \vDash_{\pi_2} \neg f \text{ and } b \wedge r \nvDash_{\pi_3} \neg f, \, b \wedge r \nvDash_{\pi_3} f.$$

Clearly, in the previous example the possibility distributions π_2 and π_3 are not desirable. It means that another constraint must be added in order to select a subset of $\Pi(\Delta, W)$. To this aim, an interpretation ω of Ω is viewed as a pair of conjuncts $\omega = x \wedge y$ where x is an interpretation only constructed from propositional symbols appearing in Δ or W while y is an interpretation only constructed from propositional symbols which neither appear in Δ nor in W. Then we can state that:

Definition 3: A possibility distribution π of $\Pi(\Delta, W)$ is said to cope with irrelevance w.r.t. (Δ, W) iff for all interpretations $\omega = x \wedge y$ and $\omega' = x' \wedge y'$ we have:

$$\text{if } x = x' \text{ then } \pi(\omega) = \pi(\omega').$$

Clearly such a π satisfies $\Pi(x) = \max_y \pi(x \wedge y) = \Pi(x \wedge y) \, \forall y$. We denote by $\Pi_R(\Delta, W)$ the set of possibility distributions of $\Pi(\Delta, W)$ which cope with irrelevance w.r.t. (Δ, W). The new inference relation, denoted by $\vDash_{\forall \Pi R}$, is defined as

$$(\Delta, W) \vDash_{\forall \Pi R} \alpha \to \beta \quad \text{iff} \quad \forall \pi \in \Pi_R(\Delta, W), \, \alpha \vDash_\pi \beta.$$

Example 3 (continued): Let us consider the Example 3 where $\Delta = \{b \to f\}$, and:

$$\Omega = \{ \, \omega_0: \neg b \wedge \neg f \wedge \neg r, \, \omega_1: \neg b \wedge \neg f \wedge r, \, \omega_2: \neg b \wedge f \wedge \neg r, \, \omega_3: \neg b \wedge f \wedge r,$$
$$\omega_4: b \wedge \neg f \wedge \neg r, \, \omega_5: b \wedge \neg f \wedge r, \, \omega_6: b \wedge f \wedge \neg r, \, \omega_7: b \wedge f \wedge r \}$$

Each π of $\Pi(\Delta, W)$ satisfies:

$$C'_1: \max(\pi(\omega_6), \pi(\omega_7)) > \max(\pi(\omega_4), \pi(\omega_5)).$$

Assume now that $(\Delta, W) \nvDash_{\forall \Pi R} b \wedge r \to f$, this means that there exists at least one possibility distribution π such that:

$$C'_2: \pi(\omega_5) \geq \pi(\omega_7).$$

This implies (using C'_1) that $\pi(\omega_6) > \pi(\omega_7)$, and this contradicts the fact that π is relevant to (Δ, W), since $\omega_6 = (x = b \wedge f, y = \neg r)$, $\omega_7 = (x' = b \wedge f, y' = r)$ and $x = x'$. Consequently:

$$(\Delta, W) \vDash_{\forall \Pi R} b \wedge r \to f. \quad\quad\quad \blacksquare$$

3.3. Consistency as an Irrelevance Property

There is another situation where we want to jump to some conclusions that the universal possibilistic consequence relation does not produce. This situation is encountered when we have an observation α which is consistent with our beliefs (in a classical sense), and thus we would like to recover all the classical entailments obtained from $\{\alpha\} \cup \Delta^* \cup W$, where Δ^* is the set of formulas obtained by turning rules in Δ into strict rules. This situation is illustrated by the following example:

Example 4: Let us consider two default rules $\Delta = \{b \to f, l \to w\}$, where the second rule reads "generally, if one has legs, one walks". Intuitively, given a bird having legs we like to deduce that it flies. The universal possibilistic consequence relation cannot infer such conclusion. Indeed, let:

$$\Omega = \{\ \omega_0: \neg b \wedge \neg f \wedge \neg l \wedge w,\ \omega_1: \neg b \wedge \neg f \wedge l \wedge w,\ \omega_2: \neg b \wedge f \wedge \neg l \wedge w,$$
$$\omega_3: \neg b \wedge f \wedge l \wedge w,\ \omega_4: b \wedge \neg f \wedge \neg l \wedge w,\ \omega_5: b \wedge \neg f \wedge l \wedge w,$$
$$\omega_6: b \wedge f \wedge \neg l \wedge w,\ \omega_7: b \wedge f \wedge l \wedge w,\ \omega_8: \neg b \wedge \neg f \wedge \neg l \wedge \neg w,$$
$$\omega_9: \neg b \wedge \neg f \wedge l \wedge \neg w,\ \omega_{10}: \neg b \wedge f \wedge \neg l \wedge \neg w,\ \omega_{11}: \neg b \wedge f \wedge l \wedge \neg w,$$
$$\omega_{12}: b \wedge \neg f \wedge \neg l \wedge \neg w,\ \omega_{13}: b \wedge \neg f \wedge l \wedge \neg w,\ \omega_{14}: b \wedge f \wedge \neg l \wedge \neg w,$$
$$\omega_{15}: b \wedge f \wedge l \wedge \neg w\ \}$$

Each possibility distribution π of (Δ, W) satisfies:

$$\Pi(b \wedge f) > \Pi(b \wedge \neg f)$$
$$\Pi(l \wedge w) > \Pi(l \wedge \neg w)$$

These constraints induce the set of constraints C' on models:

C'$_1$: $\max(\pi(\omega_6), \pi(\omega_7), \pi(\omega_{14}), \pi(\omega_{15})) > \max(\pi(\omega_4), \pi(\omega_5), \pi(\omega_{12}), \pi(\omega_{13}))$.
C'$_2$: $\max(\pi(\omega_1), \pi(\omega_3), \pi(\omega_5), \pi(\omega_7)) > \max(\pi(\omega_9), \pi(\omega_{11}), \pi(\omega_{13}), \pi(\omega_{15}))$.

We can easily check that the following possibility distribution:

$$\pi(\omega_6) = 1\ ;\ \pi(\omega_5) < 1, \text{ and } \pi(\text{otherwise}) < \pi(\omega_5)$$

is compatible with (Δ, W) but $b \wedge l \models_\pi \neg f$.   ∎

Let us restrict again the class of possibility distributions compatible with (Δ, W):

Definition 4: A possibility distribution π of $\Pi_{RI}(\Delta, W)$ is said to be π-consistent with (Δ, W) iff for each model of $W \cup \Delta^*$ we have $\pi(\omega) = 1$, and $\pi(\omega) < 1$ otherwise.

The idea is that interpretations of $W \cup \Delta^*$ should have the maximum degree of possibility since no restriction bears on them. We denote by $\Pi_{RC}(\Delta, W)$ a set of possibility distributions of $\Pi_R(\Delta, W)$ which are π-consistent with (Δ, W). The inference relation, denoted by $\models_{\forall \Pi RC}$, is defined as

$$(\Delta, W) \models_{\forall \Pi RC} \alpha \to \beta \quad \text{iff} \quad \forall \pi \in \Pi_{RC}(\Delta, W), \alpha \models_\pi \beta.$$

Example 4 (continued): Let us consider $\Delta = \{b \rightarrow f, 1 \rightarrow w\}$. Then $(\Delta, W) \vDash_{\forall\prod RC}$ $b \wedge 1 \rightarrow f$. Indeed, each possibility distribution π which is compatible, coping with irrelevance and π-consistent with (W, Δ) must satisfies: C'_1, C'_2 and C'_3: $\pi(\omega) = 1$ for and only for $\omega \in \{\omega_0, \omega_1, \omega_2, \omega_3, \omega_6, \omega_7, \omega_8, \omega_{10}, \omega_{14}\}$. The first two constraints become:

$$C'_1: 1 >_\pi \max\{\omega_4, \omega_5, \omega_{12}, \omega_{13}\},$$
$$C'_2: 1 >_\pi \max\{\omega_9, \omega_{11}, \omega_{13}, \omega_{15}\}.$$

Clearly $(\Delta, W) \vDash_{\forall\prod RC} b \wedge 1 \rightarrow f$, since $\max\{\omega_7, \omega_{15}\} >_\pi \max\{\omega_5, \omega_{13}\}$. ∎

In the general case, we have:

Proposition 2: Let α be a formula which is consistent with $\Delta^* \cup W$. Then:

$$\{\alpha\} \cup \Delta^* \cup W \vdash \beta \text{ iff } (\Delta, W) \vDash_{\forall\prod RC} \alpha \rightarrow \beta.$$

Using Proposition 2, it is now possible to apply transitivity to defaults when no inconsistency in the classical sense takes place. For instance, letting $\Delta = \{b \rightarrow fo, fo \rightarrow f\}$ where fo is for flying objects, it is now possible to deduce that birds fly. However, the inference $\vDash_{\forall\prod RC}$ is still cautious. Indeed, let us consider a base containing two independent triangle examples: $\Delta = \{a \rightarrow b, b \rightarrow c, a \rightarrow \neg c, x \rightarrow y, y \rightarrow z, x \rightarrow \neg z\}$. Let π be defined as:

$\pi(\omega) = 1$ for any ω which is a model of $\Delta^* \cup W$

(hence the consistency condition is satisfied),

$\pi(a \wedge b \wedge \neg c \wedge \neg x \wedge \neg y \wedge z) = t < 1$,
$\pi(\neg a \wedge \neg b \wedge c \wedge x \wedge y \wedge \neg z) = s < t$,
$\pi(\text{otherwise}) = r < s$.

We can check that this possibility distribution π belongs to $\prod_{RC}(\Delta, W)$, but it is not possible to deduce that "$a \wedge x$ generally implies b", namely the inference $(\Delta, W) \vDash_{\forall\prod RC}$ $x \wedge a \rightarrow b$ is not valid, since $\prod(a \wedge b \wedge x) = \prod(a \wedge \neg b \wedge x) = r$.

3.4. Generation of a Normality Ordering

A radical way of coping with the cautiousness of $\vDash_{\forall\prod}$ is to only pick one possibility distribution among those in $\prod(\Delta, W)$. We can show that this leads to add to System P, the so-called rational monotony property (Lehmann, 1989) expressed by:

Rational Monotony: from $\alpha \vdash\!\sim \delta$ then either $\alpha \vdash\!\sim \neg\beta$ or $\alpha \wedge \beta \vdash\!\sim \delta$.

The disjunctive form of rational monotony explains why for consistent beliefs (Δ, W) there are several possibility distributions in $\prod(\Delta, W)$ agreeing with rational monotony (since we are then led to consider several possible sets of constraints). It has been shown that the inference relation which can be defined from any of these possibility distributions corresponds to a superset of Δ, called a rational extension, closed under System P and the rational monotony property (see (Lehmann and Magidor, 1992) for a statement of this fact in terms of ranked models). The problem is then to find the "best" possibility distribution compatible with (Δ, W) that defines a rational extension of Δ. For this purpose, we propose to use the minimum specificity principle since it considers each interpretation as normal as possible, namely it assigns to each world ω the highest possibility level without violating the constraints. It turns out that this way of selecting a single possibility distribution is equivalent to the one used by Lehmann(1989) for defining the selected rational extension, called *the rational closure*.

In this section, we describe an algorithm which picks the possibility distribution obtained by using the minimum specificity principle. For each possibility distribution π, we associate its qualitative counterpart, denoted by $>_\pi$, defined by $\omega >_\pi \omega'$ if and only if $\pi(\omega) > \pi(\omega')$, which can be viewed as a well-ordered partition[2] $(E_1, ..., E_m, E_{m+1})$ of Ω such that:

$$\forall\, \omega \in E_i,\ \forall\, \omega' \in E_j,\qquad \pi(\omega) > \pi(\omega')\quad \text{iff}\quad i < j.$$

E_{m+1} is a subset of impossible interpretations such that $\pi(\omega) = 0$, and is therefore denoted $E_\perp$. In a similar way a complete pre-order $\geq_\pi$ is defined as: $\forall\, \omega \in E_i,\ \forall\, \omega' \in E_j$, $\omega \geq_\pi \omega'$ iff $i \leq j$ (for $i \leq m+1$, $j \geq 1$). And $\omega =_\pi \omega'$ iff $\omega \geq_\pi \omega'$ and $\omega' \geq_\pi \omega$. By convention, E_1 represents the most normal states of facts. Thus, a possibility distribution partitions Ω into classes of equally possible interpretations. Note that each possibility distribution has exactly one comparative counterpart, but for a given comparative possibility distribution $>_\pi$ admits an infinite number of representations on the unit interval. However, all the numerical counterparts produce the same set of conclusions since they induce the same ordering of Ω.

It is obvious that for each possibility distribution π of $\prod(\Delta, W)$ we have:

$$E_\perp = \{\omega\, /\, \exists\, \alpha_i \Rightarrow \beta_i \in W \text{ and } \omega \vDash \alpha_i \wedge \neg\beta_i\}$$

Let:

$$C = \{\alpha_i \wedge \beta_i >_{\prod} \alpha_i \wedge \neg\beta_i\, /\, \alpha_i \rightarrow \beta_i \in \Delta\},$$

be the set of constraints induced by Δ, which are equivalent to

$$C' = \{\max(\omega,\ \omega \vDash \alpha_i \wedge \beta_i) >_\pi \max(\omega',\ \omega' \vDash \alpha_i \wedge \neg\beta_i)\, /\, \alpha_i \rightarrow \beta_i \in \Delta\}$$

on Ω. The ordered partition of Ω associated with $>_\pi$ using the minimum specificity principle can be obtained by the following procedure (Benferhat et al., 1992):

a. $i = 0$
b. Let $E_\perp = \{\omega\, /\, \exists\, \alpha_i \Rightarrow \beta_i \in W \text{ and } \omega \vDash \alpha_i \wedge \neg\beta_i\}$; let $\Omega = \Omega - E_\perp$.
c. While Ω is not empty repeat c.1.-c.4.:
    c.1. $i \leftarrow i + 1$
    c.2.   Put in E_i every model which does not appear in the right side of any constraints of C',
    c.3.   Remove the elements of E_i from Ω,
    c.4.   Remove from C' any constraint containing elements of E_i.

---

[2] i.e., $\Omega = E_1 \cup ... \cup E_m \cup E_{m+1}$, and for $i \neq j$ we have $E_i \cap E_j = \varnothing$.

We can show that the partition $(E_1, ..., E_m, E_\perp)$ of Ω given by the previous algorithm is unique. Let us apply now this algorithm to our penguin example $\Delta = \{b \to f, p \to \neg f\}$, and $W = \{p \Rightarrow b\}$. Let:

$$\Omega = \{\omega_0: \neg b \wedge \neg f \wedge \neg p, \ \omega_1: \neg b \wedge \neg f \wedge p, \ \omega_2: \neg b \wedge f \wedge \neg p, \ \omega_3: \neg b \wedge f \wedge p,$$
$$\omega_4: b \wedge \neg f \wedge \neg p, \ \omega_5: b \wedge \neg f \wedge p, \ \omega_6: b \wedge f \wedge \neg p, \ \omega_7: b \wedge f \wedge p\}$$

Recall that the constraints on π induced by (Δ, W) are:

$$E_\perp = \{\omega_1, \omega_3\},$$
$$C'_1: \max(\pi(\omega_6), \pi(\omega_7)) > \max(\pi(\omega_4), \pi(\omega_5)), \text{ and}$$
$$C'_2: \pi(\omega_5) > \pi(\omega_7).$$

The models which do not appear in the right side of any constraint of C' are $\{\omega_0, \omega_2, \omega_6\}$, we call this set E_1. We remove the elements of E_1 from Ω and we remove the constraint C'_1 from C' (since assigning to ω_6 the highest possibility level makes the constraint C'_1 always satisfied). We start again the procedure and we find successively the two following sets $E_2 = \{\omega_4, \omega_5\}$ and $E_3 = \{\omega_7\}$. Finally, the well ordered partition of Ω is:

$$E_1 = \{\omega_0, \omega_2, \omega_6\} >_\pi E_2 = \{\omega_4, \omega_5\} >_\pi E_3 = \{\omega_7\} >_\pi E_\perp = \{\omega_1, \omega_3\}.$$

Let $(E_1, ..., E_m, E_\perp)$ be the obtained partition. A numerical counterpart to $>_\pi$ can be defined by

$$\pi(\omega) = \frac{m + 1 - i}{m} \text{ if } \omega \in E_i, \ i = 1, m$$
$$= 0 \text{ if } \omega \in E_\perp.$$

In our example we have $m = 3$ and $\pi(\omega_0) = \pi(\omega_2) = \pi(\omega_6) = 1$; $\pi(\omega_4) = \pi(\omega_5) = 2/3$; $\pi(\omega_7) = 1/3$, and $\pi(\omega_1) = \pi(\omega_3) = 0$. Note that this is purely a matter of convenience to use a numerical scale, and any other numerical counterpart such that $\pi(\omega) > \pi(\omega')$ iff $\omega >_\pi \omega'$ will work as well. Namely π is used as an ordinal scale. From this possibility distribution, we can compute for any propositional formula α its necessity degree $N(\alpha)$. For instance, $N(\neg p \vee \neg f) = \min\{1 - \pi(\omega) \mid \omega \models p \wedge f\} = \min(1 - \pi(\omega_3), 1 - \pi(\omega_7)) = 2/3$, while $N(\neg b \vee f) = \min\{1 - \pi(\omega) \mid \omega \models b \wedge \neg f\} = \min(1 - \pi(\omega_4), 1 - \pi(\omega_5)) = 1/3$ and $N(\neg p \vee b) = \min(1 - \pi(\omega_1), 1 - \pi(\omega_3)) = 1$.

Lastly, we can define an inference relation in the following way:

Definition 5: Let $>_\pi = (E_1, ..., E_m, E_\perp)$ be the obtained partition of Ω by using the minimum specificity principle, and let π be any numerical counterpart of $>_\pi$. Then β is said to be MSP-consequence (MSP: for minimum specificity principle) of α w.r.t. to (Δ, W), denoted by $(\Delta, W) \models_{MSP} \alpha \to \beta$, iff $\alpha \models_\pi \beta$.

We can check that the following implications are valid:

$$(\Delta, W) \models_{\forall \Pi} \alpha \to \beta \quad \Rightarrow (\Delta, W) \models_{\forall \Pi R} \alpha \to \beta$$
$$\Rightarrow (\Delta, W) \models_{\forall \Pi RC} \alpha \to \beta$$
$$\Rightarrow (\Delta, W) \models_{MSP} \alpha \to \beta.$$

The set of MSP-consequences of Δ are equivalent to the rational closure of Δ and to the set of plausible conclusions of System Z (Pearl, 1990). See (Benferhat et al., 1992) for a proof of the equivalence between System Z and MSP-entailment, and (Goldszmidt and Pearl, 1990) for the connection between System Z and the rational closure. It can be checked that in the counter-example given after Proposition 2, the MSP-entailment enables us to deduce that "a $\wedge$ x generally implies b", while $\models_{\Pi RC}$ fails. Besides, the MSP-entaiment can be conveniently encoded in the framework of possibilistic logic, a logic where formulas are made of pairs of classical logic formulas and certainty levels. See (Benferhat et al., 1992) for details.

3.5. Cautiousness of MSP-Entailment

A first case of cautiousness is the "blocking of property inheritance" problem. It corresponds to the case when a class is exceptional for a superclass with respect to some property, then the least specific possibility distribution does not allow to conclude anything about whether this class is normal with respects to other properties. For example, let us expand our penguin example by a fourth default expressing that "birds have legs", noted by $b \rightarrow l$. As we can check below, we cannot deduce that a penguin has legs. Indeed the set of rules $\{b \rightarrow f, b \rightarrow l, p \Rightarrow b, p \rightarrow \neg f\}$ leads to the following well-ordered partition of Ω in a 4-level stratification

$$E_1 = \{\omega_0: \neg p \wedge \neg b \wedge \neg f \wedge \neg l, \omega_1: \neg p \wedge \neg b \wedge \neg f \wedge l, \omega_2: \neg p \wedge \neg b \wedge f \wedge \neg l,$$
$$\omega_3: \neg p \wedge \neg b \wedge f \wedge l, \omega_4: \neg p \wedge b \wedge f \wedge l\}$$
$$E_2 = \{\omega_5: \neg p \wedge b \wedge \neg f \wedge \neg l, \omega_6: \neg p \wedge b \wedge \neg f \wedge l, \omega_7: \neg p \wedge b \wedge f \wedge \neg l,$$
$$\omega_8: p \wedge b \wedge \neg f \wedge l, \omega_9: p \wedge b \wedge \neg f \wedge \neg l\}$$
$$E_3 = \{\omega_{14}: p \wedge b \wedge f \wedge \neg l, \omega_{15}: p \wedge b \wedge f \wedge l\}.$$
$$E_\perp = \{\omega_{10}: p \wedge \neg b \wedge \neg f \wedge \neg l, \omega_{11}: p \wedge \neg b \wedge \neg f \wedge l, \omega_{12}: p \wedge \neg b \wedge f \wedge \neg l,$$
$$\omega_{13}: p \wedge \neg b \wedge f \wedge l\}.$$

It can be easily seen that it will not be possible to deduce that a penguin has legs since the two preferred models (where penguin is true) are ω_8 and ω_9, one where l is true and one where l is false. The fact that penguin is a non-typical subclass of birds due to the failure of the fly property leads to assume that this subclass may be exceptional in other respects. However, we would like to be able to draw the conclusion that penguins have legs since intuitively this has nothing to do with the failure of the fly property. A natural idea is then to try to express some kind of independence between the fact of having legs and the fact of flying, for birds.

A second situation of cautiousness of the MSP-entailment is illustrated by the following example $\Delta = \{s \rightarrow b, s \rightarrow t\}$ which reads "Swedes normally are blond" and "Swedes normally are tall". The minimum specificity principle generates the following partition:

$$E_1 = \{\omega_0: \neg s \wedge b \wedge t, \omega_1: \neg s \wedge \neg b \wedge t, \omega_2: \neg s \wedge \neg b \wedge \neg t, \omega_3: \neg s \wedge b \wedge \neg t, \omega_4: s \wedge b \wedge t\}$$
$$E_2 = \{\omega_5: s \wedge b \wedge \neg t, \omega_6: s \wedge \neg b \wedge t, \omega_7: s \wedge \neg b \wedge \neg t\}.$$

Note that we get the same partition as the one given by classical logic, namely E_1 corresponds to models of Δ^* and E_2 to interpretations which do not satisfy Δ^*. From this partition the MSP-entailment cannot infer that "non-tall Swedes are blonds", namely s ∧ ¬t ⊨$_{MSP}$ b is not valid. Here again the cautiousness of MSP-entailment can be overcome by adding expressing independence between "being tall" and "being blond". This issue is discussed in next Section.

4. In Search of Independence

When the MSP-entailment still appears too cautious, we are looking for a possibility theory-based encoding of pieces of information of the form "in the context α, δ has no influence on β" (equivalently, "in the context α, β is independent of δ (or irrelevant to δ)"). We denote this independence information by $I(\delta, \alpha \rightarrow \beta)$. The meaning of this statement is that in the context α, learning δ does not change our belief about β.

4.1. Possibilistic Independence

A natural way for expressing independence, which takes its inspiration from the probabilistic setting, in terms of conditional necessity is the following. Accepting β is not affected by δ in the context α iff

$$N(\beta \mid \alpha) = N(\beta \mid \alpha \wedge \delta) > 0. \qquad \text{(A)}$$

When (A) is satisfied, we say that β is *absolutely independent* of δ in the context α (Benferhat et al., 1994). Note that (A) can be written $\Pi(\neg\beta \mid \alpha) = \Pi(\neg\beta \mid \alpha \wedge \delta) < 1$. It is not related to the equality $\Pi(\beta \mid \alpha) = \Pi(\beta \mid \alpha \wedge \delta)$ which is similar to the probabilistic condition of independence in terms of possibility measures. Moreover, strong independence implies the so-called unrelatedness proposed by Nahmias(1978) and defined as: "β is unrelated to δ in the context α if and only if $\Pi(\delta \wedge \beta \mid \alpha) = \min(\Pi(\beta \mid \alpha), \Pi(\delta \mid \alpha))$". The converse however does not hold. It is easy to verify that (A) is equivalent to the set of three inequalities (Dubois et al., 1994)

$$\Pi(\alpha \wedge \beta) > \Pi(\alpha \wedge \neg\beta);$$
$$\Pi(\alpha \wedge \beta \wedge \delta) > \Pi(\alpha \wedge \neg\beta \wedge \delta);$$
$$\Pi(\alpha \wedge \delta \wedge \neg\beta) \geq \Pi(\alpha \wedge \neg\delta \wedge \neg\beta).$$

It means that "generally if α is true then β is true", "generally, if $\alpha \wedge \delta$ is true then β is true", and it is forbidden to claim that when $\neg\beta \wedge \alpha$ is true then δ is false. The latter condition appears to be too restrictive in the setting of exception-tolerant generic rules since claiming that δ does not affect our acceptance of β in the context α should leave us free to accept δ or not when β is false. Note that the third condition corresponds to the negation of the default rule $\neg\beta \wedge \alpha \rightarrow \neg\delta$ which is not a constraint of the same nature as before. Lastly, the equality used in (A) is not in the spirit of possibility theory, since what is important in possibility theory is only the ordering and not the numbers themselves.

A less restrictive condition of independence is the following: δ does not question the acceptance of β in the context α iff (Dubois et al., 1994)

$$N(\beta \mid \alpha) > 0 \text{ and } N(\beta \mid \alpha \wedge \delta) > 0. \qquad \text{(B)}$$

When (B) is satisfied we say that β is *qualitatively independent* of δ in the context α. Clearly, absolute independence implies qualitative independence (namely, (A) entails (B)). Note that (B) is sensitive to negation, i.e., we cannot change δ into $\neg\delta$ in (B), contrarily to probabilistic independence. Similarly, β and δ do not play symmetric roles as is the case in probability theory, i.e., (B) is not equivalent to $N(\delta \mid \alpha) > 0$ and $N(\delta \mid \alpha \wedge \beta) > 0$. Hence if we want to express that β is accepted whether δ is true or false in the context α we have to explicitly state that $N(\beta \mid \alpha \wedge \delta) > 0$ and $N(\beta \mid \alpha \wedge \neg\delta) > 0$ (which then implies $N(\beta \mid \alpha) > 0$). The condition (B) is also equivalent to

$$\Pi(\beta \wedge \alpha) > \Pi(\neg\beta \wedge \alpha) \text{ and } \Pi(\beta \wedge \alpha \wedge \delta) > \Pi(\neg\beta \wedge \alpha \wedge \delta).$$

It expresses an invariance property of the ordering of plausibility whether or not we focus on δ-worlds. In the context of conditional knowledge bases, we are interested in using qualitative independence of δ with respect to $\alpha \to \beta$ only if $\alpha \to \beta \in \Delta$. Namely, declaring that δ is qualitatively independent of $\alpha \to \beta$ means that $\alpha \to \beta \in \Delta$ implies $\alpha \wedge \delta \to \beta \in \Delta$ as well (the latter being $N(\beta \mid \alpha) > 0 \Rightarrow N(\beta \mid \alpha \wedge \delta) > 0$, as used in (Benferhat et al., 1994)). In other words, assuming that the default $\alpha \to \beta$ is independent of the truth of δ just amounts to augmenting Δ with one more specific default, namely $\alpha \wedge \delta \to \beta$ if $\alpha \to \beta \in \Delta$, or if $\alpha \to \beta$ can be deduced from Δ in a sense to be made precise (see Section 4.2). Two defaults $\alpha \wedge \delta \to \beta$ and $\alpha \wedge \neg\delta \to \beta$ must be added if $\alpha \to \beta$ is claimed to hold regardless of the truth or falsity of δ.

There may be one objection against the definition of qualitative independence as the implication $N(\beta \mid \alpha) > 0 \Rightarrow N(\beta \mid \alpha \wedge \delta) > 0$. It considers that everything is independent of a given formula δ if $N(\beta \mid \alpha) = 0$. The following equivalence-based definition of independence (called *C-independence* for short) remedies this drawback: δ is independent of β in the context α iff

$$N(\beta \mid \alpha) > 0 \text{ iff } N(\beta \mid \alpha \wedge \delta) > 0. \qquad (C)$$

Note that (C) is equivalent to the disjunction of (B) and $N(\beta \mid \alpha \wedge \delta) = N(\beta \mid \alpha) = 0$. However $N(\beta \mid \alpha \wedge \delta) = N(\beta \mid \alpha) = 0$ includes the case when $N(\neg\beta \mid \alpha \wedge \delta) > 0$ and $N(\neg\beta \mid \alpha) = 0$ and the one when $N(\neg\beta \mid \alpha \wedge \delta) = 0$ and $N(\neg\beta \mid \alpha) > 0$. The former case means that δ is a reason to reject β in the context α, and the latter case indicates that δ cancels our belief in $\neg\beta$ in the context α. So (C) is a rather complex notion. In (Dubois et al., 1996) two forms of irrelevance are considered: (B) for accepted beliefs and the situation where $N(\beta \mid \alpha \wedge \delta) = N(\beta \mid \alpha) = N(\neg\beta \mid \alpha \wedge \delta) = N(\neg\beta \mid \alpha) = 0$, called uninformativeness, because in that case δ does not inform at all about β or its negation in the context α.

4.2. Application to Plausible Reasoning

We now present a general procedure for handling default rules and independence together. Let $\mathcal{I} = \{I(\delta_i, \alpha_i \to \beta_i) / i = 1, m\}$ be a set of pieces of independence information.

Definition 7: A possibility distribution π of $\Pi(\Delta, W)$ is said to satisfy the set $\mathcal{I}$ iff for each independence information $I(\delta_i, \alpha_i \to \beta_i)$ of $\mathcal{I}$ we have: if $\Pi(\beta_i \wedge \alpha_i) > \Pi(\neg\beta_i \wedge \alpha_i)$ then $\Pi(\beta_i \wedge \alpha_i \wedge \delta_i) > \Pi(\neg\beta_i \wedge \alpha_i \wedge \delta_i)$, where Π is defined from π.

We denote $\Pi_\mathcal{I}(\Delta, W)$ the set of π's in $\Pi(\Delta, W)$ which satisfy $\mathcal{I}$. The following algorithm computes the possibility distribution which is in $\Pi_\mathcal{I}(\Delta, W)$ and which computes a possibility distribution in $\Pi(\Delta, W)$ taking into account independence constraints. We denote by: $\mathcal{C}_\Delta = \{\max(\omega \mid \omega \models \alpha_i \wedge \beta_i) >_\pi \max(\omega \mid \omega \models \alpha_i \wedge \neg\beta_i) \mid \alpha_i \to \beta_i \in \Delta\}$ the set of constraints on models induced by Δ, by $D(\mathcal{C}_\Delta)$ the set of interpretations which do not appear in the right side of any constraint in $\mathcal{C}_\Delta$. The algorithm is:

a. $i := 0$;

b. Let $E_\perp = \{\omega \mid \exists\ \alpha_j \Rightarrow \beta_j \in W \text{ s.t. } \omega \models \alpha_j \wedge \neg\beta_j\}$; $\Omega := \Omega - E_\perp$.

c. While $\Omega \neq \emptyset$ do c.1.-c.5

    c.1. $i := i + 1$; $E_i := \{\omega\ /\ \omega \in \Omega \text{ and } \omega \in D(\mathbb{C}_\Delta)\}$;

    c.2.   Remove from $\mathcal{I}$ each $I(\delta_i, \alpha_i \to \beta_i)$ s.t., $\exists\ \omega \in E_i$ and $\omega \models \neg\beta_i \wedge \alpha_i$

    c.3.   For each $I(\delta_i, \alpha_i \to \beta_i)$ of $\mathcal{I}$ s.t., $\exists\ \omega \in E_i$ and $\omega \models \beta_i \wedge \alpha_i$

        c.3.1. $\mathbb{C}_\Delta = \mathbb{C}_\Delta \cup \{\max(\omega \mid \omega \models \alpha_i \wedge \beta_i \wedge \delta_i) >_\pi \max(\omega \mid \omega \models \alpha_i \wedge \neg\beta_i \wedge \delta_i)\}$

        c.3.2. $\mathcal{I} = \mathcal{I} - \{I(\delta_i, \alpha_i \to \beta_i)\}$

    c.4.   If $E_i = \emptyset$ Then Stop (inconsistent beliefs)

    c.5.   Remove from $\mathbb{C}_\Delta$ any constraint containing at least one interpretation of E_i

    c. 6. $\Omega := \Omega - E_i$

d. Return $E_1, ..., E_i, E_\perp$.

Let us briefly explain the main ideas of this algorihtm. Step b simply puts in $E_\perp$ the set of interpretations which falsifies at least one strict rule. Step c.1. put in a stratum E_i the interpretations which do not appear in the right side of any constraint in $\mathbb{C}_\Delta$, namely we try to give to the interpretations the highest possible values without violating any constraint of $\mathbb{C}_\Delta$ (this is the idea of the minimum specificity principle). Step c.2. removes any independence information $I(\delta_i, \alpha_i \to \beta_i)$ s.t., $\exists\ \omega \in E_i$ and $\omega \models \neg\beta_i \wedge \alpha_i$. This is due to the fact $\alpha_i \to \beta_i$ will not hold in the constructed possibility distribution. In the case where there is an interpretation in E_i s.t. $\omega \models \beta_i \wedge \alpha_i$ for some $I(\delta_i, \alpha_i \to \beta_i)$, the constraint $\max\{\omega \mid \omega \models \alpha_i \wedge \beta_i \wedge \delta_i\} >_\pi \max\{\omega \mid \omega \models \alpha_i \wedge \neg\beta_i \wedge \delta_i\}$ is added to $\mathbb{C}_\Delta$. This means that $\alpha_i \to \beta_i$ will hold in the constructed possibility distribution, and therefore the added constraint guarantees also that the default $\alpha_i \wedge \delta_i \to \beta_i$ will also hold in the constructed possibility distribution. Step c.4. indicates a case of inconsistency which can be due either to the fact that (Δ, W) is inconsistent, or to the fact that $\mathcal{I}$ contradicts (Δ, W). For instance, we can easily check that $\mathcal{I} = \{I(p, b \to f)\}$ contradicts $\Delta = \{b \to f, p \wedge b \to \neg f\}$. In this algorithm, we simply detect the inconsistency but we do not solve it. Lastly, Step c.5. removes from $\mathbb{C}_\Delta$ the satisfied constraints, namely the ones which contain at least one interpretation of E_i.

Let π be the possibility distribution computed by the algorithm (Step d). Then we can check that $(\Delta, W) \models_{\forall\Pi} \alpha \to \beta$ implies $\alpha \models_\pi \beta$. Moreover, it can be shown that there is no distributions in $\Pi(\Delta, W)$ which satisfy the set of constraints in $\mathcal{I}$ (in the sense of Definition 7), and which are less specific than π.

Let us go back to the penguin example containing the following rules $\Delta = \{b \to f, b \to l, p \to \neg f\}$, $W = \{p \Rightarrow b\}$ and given in Section 3.5.1. We have:

$\Omega = \{\omega_0: \neg p \wedge \neg b \wedge \neg f \wedge \neg l, \omega_1: \neg p \wedge \neg b \wedge \neg f \wedge l, \omega_2: \neg p \wedge \neg b \wedge f \wedge \neg l,$

    $\omega_3: \neg p \wedge \neg b \wedge f \wedge l, \omega_4: \neg p \wedge b \wedge f \wedge l, \omega_5: \neg p \wedge b \wedge \neg f \wedge \neg l, \omega_6: \neg p \wedge b \wedge \neg f \wedge l,$

    $\omega_7: \neg p \wedge b \wedge f \wedge \neg l, \omega_8: p \wedge b \wedge \neg f \wedge l, \omega_9: p \wedge b \wedge \neg f \wedge \neg l, \omega_{10}: p \wedge \neg b \wedge \neg f \wedge \neg l,$

    $\omega_{11}: p \wedge \neg b \wedge \neg f \wedge l, \omega_{12}: p \wedge \neg b \wedge f \wedge \neg l, \omega_{13}: p \wedge \neg b \wedge f \wedge l, \omega_{14}: p \wedge b \wedge f \wedge \neg l,$

    $\omega_{15}: p \wedge b \wedge f \wedge l\}$.

The constraints $\mathcal{C}_\Delta$ are:

C_1: $\max(\omega_4, \omega_7, \omega_{14}, \omega_{15}) > \max(\omega_5, \omega_6, \omega_8, \omega_9)$ (corresponding to b→f)
C_2: $\max(\omega_4, \omega_6, \omega_8, \omega_{15}) > \max(\omega_5, \omega_7, \omega_9, \omega_{14})$ (corresponding to b→l)
C_3: $\max(\omega_8, \omega_9, \omega_{10}, \omega_{11}) > \max(\omega_{12}, \omega_{13}, \omega_{14}, \omega_{15})$ (corresponding to p→¬f)

Let us assume that we have the following four pieces of independence information: $\mathcal{I} =$ {I(¬f, b → l), I(p, b → l), I(f, b → l), I(¬p, b → l)}.
Let us now apply the algorithm given above in order to get the least specific possibility distribution which satisfies $\mathcal{I}$ and which is compatible with (Δ, W). Step b leads to put:

$$E_\bot = \{\omega_{10}, \omega_{11}, \omega_{12}, \omega_{13}\}.$$

The constraint C_3 becomes:

$$C'_3\colon \max(\omega_8, \omega_9) > \max(\omega_{14}, \omega_{15})$$

Step c.1 leads to have: $E_1 = \{\omega_0, \omega_1, \omega_2, \omega_3, \omega_4\}$.
Note that the interpretation ω_4 satisfies b ∧ l but none of $\{\omega_0, \omega_1, \omega_2, \omega_3, \omega_4\}$ satisfies b ∧ ¬l (Step c.2. does not remove anything from $\mathcal{I}$), hence Step c.3. says that the following four constraints must be added to $\mathcal{C}_\Delta$:

C_4: $\max(\omega_6, \omega_8) > \max(\omega_5, \omega_9)$ (corresponding to I(¬f, b → l))
C_5: $\max(\omega_8, \omega_{15}) > \max(\omega_9, \omega_{14})$ (corresponding to I(p, b → l))
C_6: $\max(\omega_4, \omega_{15}) > \max(\omega_7, \omega_{14})$ (corresponding to I(f, b → l))
C_7: $\max(\omega_4, \omega_6) > \max(\omega_5, \omega_7)$ (corresponding to I(¬p, b → l))

Step c.5. removes from C_1, C_2, C_6 and C_7 from $\mathcal{C}_\Delta$, since they contain the interpretation ω_4 which is in E_1. At the end of Step c.6., $\mathcal{I} = \varnothing$ and $\mathcal{C}_\Delta = \{C'_3, C_4, C_5\}$. Let us repeat again the algorithm, we get: $E_2 = \{\omega_6, \omega_7, \omega_8\}$. All the remaining constraints are satisfied, and hence we get: $E_3 = \{\omega_5, \omega_{14}, \omega_{15}, \omega_9\}$.

It is now possible to conclude that penguins have legs since from this partition ω_8 has a higher plausibility degree than ω_9.
Lastly, in the Swedish example given in Section 3.5.2., from the default base {s → b, s → t} and the independence information "in the context s, b is independent of t or ¬t", we can check that applying the above algorithm leads to infer that b still follows from s ∧ ¬t.

5. Semantic and syntactic fusion
In this section, we propose syntactic and semantic approaches for fusing pieces of uncertain information provided by different sources. There exists a large panoply of operations which have been studied for aggregating fuzzy sets (e.g., Dubois and Prade, 1985). They can be used for combining possibility distribution π_i's whose valuation scales are commensurate. This section provides the syntactic counterpart of classical fuzzy set aggregation operations on the possibilistic beliefs bases $\Sigma_{\pi i}$ associated with the π_i's.
In the following, we first describe the syntactic counterpart of the combination of two possibility distributions by an operator ⊕ which is very weakly constrained. The extension

to more than two possibility distributions is straightforward, see (Benferhat et al., 1997b). Then, we propose to discuss some particular cases of $\oplus$ which are of special interest.

5.1. Syntactic counterpart of the combination of two possibility distributions

Let Σ_1 and Σ_2 be two possibilistic belief bases. Let π_1 and π_2 be their associated possibility distributions. Let $\oplus$ be a semantic operator which aggregate the two possibility distributions π_1 and π_2 into a new one $\pi_\oplus$. Then from Σ_1 and Σ_2, we are interested in building a new possibilistic belief bases $\Sigma_\oplus$ such that $\pi_{\Sigma_\oplus} = \pi_\oplus$.

We first analyse the general case where $\oplus$ is very weakly constrained. Then we discuss some interesting combination modes in next section. The only requirements for $\oplus$ are the following properties:

$$1 \oplus 1 = 1;$$

if $a \geq c$, $b \geq d$ then $a \oplus b \geq c \oplus d$ (monotonicity).

The first one acknowledges the fact that if two sources agree for considering that ω is fully possible, the result of the combination should confirm it. The second property expresses that a possibility degree resulting from a combination cannot decrease if the combined degrees increase. Let us first consider the syntactic counterpart of $\oplus$ (denoted by the same symbol) when the combination is applied to two one-formula belief bases $\Sigma_1 = \{(\phi\ a)\}$ and $\Sigma_2 = \{(\psi\ b)\}$. Then we can easily check that this results into three possibilistic formulas:

Lemma 1: $\Sigma_\oplus = \Sigma_1 \oplus \Sigma_2$
$$= \{(\phi, 1-(1-a)\oplus 1)\} \cup \{(\psi, 1-1\oplus(1-b))\} \cup \{(\phi \vee \psi, 1-(1-a)\oplus(1-b))\}.$$

The result given in Lemma 1 can be generalized to the case of general possibilistic belief bases. Let $\Sigma_1 = \{(\phi_i, a_i) : i \in I\}$ and $\Sigma_2 = \{(\psi_j, b_j) : j \in J\}$. Namely, we have the following result.

Proposition 3: Let $\pi_\oplus$ be the result of the combination of π_1 and π_2 based on the operator $\oplus$. Then, $\pi_\oplus$ is associated with the following belief base:

$$\Sigma_\oplus = \{(\phi_i\ 1-(1-a_i)\oplus 1) : (\phi_i\ a_i) \in \Sigma_1\} \cup$$
$$\{(\psi_j\ 1-1\oplus(1-b_j)) : (\psi_j\ b_j) \in \Sigma_2\} \cup$$
$$\{(\phi_i \vee \psi_j\ 1-(1-a_i)\oplus(1-b_j)) : (\phi_i\ a_i) \in \Sigma_1 \text{ and } (\psi_j\ b_j) \in \Sigma_2\}.$$

The proof is given in Appendix. Next sub-sections discuss particular cases of the combination operator $\oplus$ which are meaningful. The first one (idempotent conjunction) is meaningful when the sources are consistent but dependent, the second one (idempotent disjunction) is appropriate when the sources are highly conflicting, the third one deals with independent sources, and the last one is the usual weighted average.

5.2. Idempotent conjunction

The first combination mode that we consider in this section is the idempotent conjunction (i.e., the minimum) of possibility distributions. Namely define:

$$\forall \omega, \pi_{cm}(\omega) = \min(\pi_1(\omega), \pi_1(\omega)). \tag{CM}$$

Conjunctive aggregation makes sense if all the sources are regarded as equally and fully reliable since any value that is considered as impossible by one source but possible by the others are rejected. This is true for any combination operation $\oplus$ such that $a \oplus b \leq \min(a,b)$, $\forall a,b$. The operation min is the greatest and the only idempotent conjunction function. To clarify this point of view, let us assume that we only have two binary valued possibility distributions π_1 and π_2 such that each π_i partitions the set of classical interpretations into two subsets, namely A_i containing the completely possible interpretations (i.e., $\forall \omega \in A_i$, $\pi_i(\omega)=1$) and $\Omega-A_i$ containing the completely impossible interpretations (i.e., $\forall \omega \in \Omega-A_i$, $\pi_i(\omega)=0$). The result of the combination of π_i using (CM) leads to partition Ω into two subsets $(A_1 \cap A_2, \Omega-(A_1 \cap A_2))$. The conjunction mode (CM) in this case is natural if $A_1 \cap A_2$ is not empty.

An important issue with conjunctive combination as defined by (CM) is the fact that the result may be subnormalized, i.e., it may happen that $\nexists \omega$, $\pi_{cm}(\omega)=1$. In that case it expresses a conflict between the sources. Clearly the conjunctive mode makes sense if all the π_i's significantly overlap, i.e., $\exists \omega$, $\forall i$, $\pi_i(\omega)=1$, expressing that there is at least a value of ω that all sources consider as completely possible. If $\forall \omega$, $\pi_{cm}(\omega)$ is significantly smaller than 1 this mode of combination is debatable since in that case at least one of the sources (or experts) is likely to be wrong.

Besides, if two sources provide the same information $\pi_1=\pi_2$, the result of the conjunction is still the same distribution with $\oplus=\min$. More generally, if the information provided by a source k is more specific than the information given by the other then $\pi_{cm}=\pi_k$.

Syntactic counterpart

Letting $\oplus=\min$ in Proposition 3, the min-based conjunction mode (CM) leads simply to take the union of Σ_1 and Σ_2 at the syntactic level, namely:

Proposition 4: $\Sigma_{cm} = \Sigma_1 \cup \Sigma_2.$

5.3. Idempotent disjunction

When the pieces of information provided by the sources are highly conflicting (i.e., If $\forall \omega$, $\pi_{cm}(\omega)$ is significantly smaller than 1), then the disjunction (i.e., the maximum) of possibility distributions can be more appropriate. Namely define:

$$\forall \omega, \pi_{dm}(\omega) = \max(\pi_1(\omega), \pi_2(\omega)). \qquad \text{(DM)}$$

The disjunctive aggregation corresponds to a weaker reliability hypothesis. Namely, in the group of sources there is at least one reliable source for sure, but we do not know which one. When we only have two binary valued possibility distributions π_1 and π_2 such that each π_i partitions the set of classical interpretations into two subsets, namely A_i containing the completely possible interpretations (i.e., $\forall \omega \in A_i$, $\pi_i(\omega)=1$) and $\Omega-A_i$ containing the completely impossible interpretations (i.e., $\forall \omega \in \Omega-A_i$, $\pi_i(\omega)=0$). The result of the combination of π_i using (DM) leads to partition Ω into two subsets $(A_1 \cup A_2, \Omega-(A_1 \cup A_2))$ respectively. The disjunction mode (DM) in this case is natural if $A_1 \cap A_2$ is empty. Besides, if two sources provide the same information $\pi_1=\pi_2$, the result of the disjunctive combination is still the same distribution. And if the information provided by a source k is less specific than the information given by the other(s) then $\pi_{dm}=\pi_k$.

Syntactic counterpart
Letting $\oplus$=max in Proposition 3, the max-based conjunction mode (DM) leads to:

Proposition 5: $\Sigma_{dm} = \{(\phi_i \vee \psi_j \min(a_i,b_j)) : (\phi_i\, a_i)\in \Sigma_1 \text{ and } (\psi_j\, b_j)\in \Sigma_2\}.$

Note that Σ_{dm} is always consistent (provided that Σ_1 or Σ_2 is consistent).

5.4. Reinforcement combination modes
The min-based conjunctive combination mode has no reinforcement effect. Namely, if source 1 assigns possibility $\pi_1(\omega)<1$ to an interpretation ω, and source 2 assigns possibility $\pi_2(\omega)<1$ to this interpretation, then overall, in the conjunctive mode, $\pi_\oplus(\omega)=\pi_1(\omega)$ if $\pi_1(\omega)<\pi_2(\omega)$, regardless of the value of $\pi_2(\omega)$. However since both sources consider ω as rather impossible, and if these opinions are independent, it may sound reasonable to consider ω as less possible than what each of the sources claims. More generally, if a pool of independent sources is divided into two unequal groups that disagree, we may want to favor the opinion of the biggest group. This type of combination cannot be modelled by the minimum operation, nor by any idempotent operation (in particular a similar argumentation holds for the disjunctive combination using maximum). What is needed is a reinforcement effect. A reinforcement effect can be obtained using a conjunctive operation other than the minimum operator. The most usual ones are the product and the so-called "Lukasiewicz t-norm" (since it is directly related to Lukasiewicz many-valued implication) :

$$\forall \omega, \ \pi_{Prod}(\omega) \ = \ \pi_1(\omega) \times \pi_2(\omega). \hspace{2cm} \text{(PROD)}$$
$$\forall \omega, \ \pi_{LUK}(\omega) \ = \ \max(0, \pi_1(\omega) + \pi_2(\omega) - 1). \hspace{1cm} \text{(LUK)}$$

These combination modes are not idempotent. Their syntactic counterpart are directly derived from Proposition 3:

Proposition 6: $\Sigma_{PROD} = \Sigma_1 \cup \Sigma_2 \cup \{(\phi_i \vee \psi_j, a_i+b_j-a_i.b_j):(\phi_i, a_i)\in \Sigma_1 \text{ and } (\psi_j, b_j)\in \Sigma_2\}$

Proposition 7: $\Sigma_{LUK}=\Sigma_1 \cup \Sigma_2 \cup$
$$\{(\phi_i \vee \psi_j, \min (1, a_i + b_j)) : (\phi_i, a_i)\in \Sigma_1 \text{ and } (\psi_j, b_j)\in \Sigma_2\}$$

5.5. Weighted Average
The last combination mode that we consider is the weighted average (WA). Let x_1 and x_2 be two non-negative real numbers such that $x_1 + x_2 =1$. Then (WA) is defined by:
$$\forall \omega, \ \pi_{WA}(\omega) \ = \ x_1 \times \pi_1(\omega) + x_2 \times \pi_2(\omega). \hspace{1.5cm} \text{(WA)}$$
The syntactic counterpart is:

Proposition 8: $\Sigma_{WA} = \{(\phi_i, x_1 \times a_i) : (\phi_i, a_i)\in \Sigma_1\} \cup \{(\psi_j, x_2 \times b_j) : (\psi_j, b_j)\in \Sigma_2\} \cup$
$$\{(\phi_i \vee \psi_j, x_1 \times a_i + x_2 \times b_j) : (\phi_i, a_i)\in \Sigma_1 \text{ and } (\psi_j, b_j)\in \Sigma_2\}$$

Note that if $x_1=1$ then:
$$\Sigma_{WA} = \{(\phi_i\ a_i) : (\phi_i\ a_i)\in \Sigma_1\} \cup \{(\phi_i \vee \psi_j\ a_i) : (\phi_i\ a_i)\in \Sigma_1 \} \equiv \ \Sigma_1 .$$

The weights x_1 and x_2 in (WA) can account for the reliability of the sources. Conjunctions (and disjunctions) can be weighted as well. For instance, the weighted version of $\min_i a_i$ is $\min_i \max(a_i, 1-w_i)$ with $\max_i w_i = 1$. The weights can be also made of functions of the amount of conflicts, or the sources can be prioritized in the combination process. See (Benferhat et al., 1997b) for more details.

6. Conclusion

Possibilistic logic is basically classical logic augmented with levels of priority or certainty. These levels may indeed reflects the degree of certainty with which propositions are held for true. Thus a possibilistic logic base is a convenient and simple way of satisfying a set of propositions according to their level of confidence. Interestingly enough such a base has a semantical counterpart under the form of an ordered set of models which have different levels of plausibility. Thus fusion of possibilistic logic bases can be equivalently performed at the semantic and at the syntactic level.

The semantics of possibilistic logic explains why defaults can be encoded as regular propositions with levels of priority. Indeed, a default rule "if p generally q" can be understood as a simple constraint expressing that in the context p, situations where q is true are more plausible than situations where q is false. Moreover, independence statements can be combined with defaults rules and encoded in the same framework. Possibilistic logic thus offers a powerful framework for reasoning under uncertainty in a qualitative way.

References

E.W. Adams (1975) The logic of conditionals. D. Reidel, Dordrecht.

C. Baral, S. Kraus, J. Minker, Subrahmanian (1992) Combining knowledge bases consisting in first order theories. Computational Intelligence, 8(1), 45-71.

S. Benferhat (1994) Handling hard rules and default rules in possibilistic logic. In: Advances in Intelligent Computing — IPMU'94, (B. Bouchon-Meunier, R.R. Yager, L.A. Zadeh, eds.), Lecture Notes in Computer Science, Vol. 945, Springer Verlag, Berlin, 1995, 302-310.

S. Benferhat, D. Dubois, H. Prade (1992) Representing default rules in possibilistic logic. Proc. of the 3rd Inter. Conf. on Principles of Knowledge Representation and Reasoning (KR'92), Cambridge, MA, Oct. 25-29, 673-684.

S. Benferhat, D. Dubois, H. Prade (1994) Expressing independence in a possibilistic framework and its application to default reasoning. Proc. of the 11th Europ. Conf. on Artificial Intelligence (ECAI'94), Amsterdam, Aug. 8-12, 150-154.

S. Benferhat, D. Dubois, H. Prade (1996a) Beyond counter-examples to nonmonotonic formalisms: A possibility-theoretic analysis. Proc. of the 12th Europ. Conf. on Artificial Intelligence (ECAI'96) (W. Wahlster, ed.), Budapest, Hungary, Aug. 11-16, Wiley, New York, 652-656.

S. Benferhat, D. Dubois, H. Prade (1996b) Coping with the limitations of rational inference in the framework of possibility theory. Proc. of the 12th Conf. on Uncertainty in Artificial Intelligence (E. Horvitz, F. Jensen, eds.), Portland, Oregon, Aug. 1-4, Morgan & Kaufmann, San Mateo, CA, 90-97.

S. Benferhat, D. Dubois, H. Prade (1997a) Nonmonotonic reasoning, conditional objects and possibility theory. Artificial Intelligence, 92, 259-276.

S. Benferhat, D. Dubois, H. Prade (1997b) From semantic to syntactic approaches to information combination in possibilistic logic. In Aggregation and Fusion of Imperfect Information, Physica Verlag, pp. 141-151.

L. Cholvy (1998) Reasoning about merging information. In Handbook of Defeasible Reasoning and Uncertainty Management Systems, Vol. 3, 233-263.

L.J. Cohen (1977) The Probable and the Provable. Clarendon Press, Oxford, UK.

D. Dubois, L. Fariñas del Cerro, A. Herzig, H. Prade (1994) An ordinal view of independence with application to plausible reasoning. Proc. of the 10th Conf. on Uncertainty in Artificial Intelligence (R. Lopez de Mantaras, D. Poole, eds.), Seattle, WA, July 29-31, 195-203

D. Dubois, L. Fariñas del Cerro, A. Herzig, H. Prade (1996) A roadmap of qualitative independence. In: Tech. Report IRIT/96-44-R, IRIT, Univ. P. Sabatier, Toulouse, France.

Dubois D., Lang J., Prade H. (1987) Theorem proving under uncertainty — A possibility theory-based approach. Proc. of the 10th Inter. Joint Conf. on Artificial Intelligence, Milano, Italy, August, 984-986.

D. Dubois, J. Lang, H. Prade (1994) Possibilistic logic. In: Handbook of Logic in Artificial Intelligence and Logic Programming, Vol. 3 (D.M. Gabbay, C.J. Hogger, J.A. Robinson, D. Nute, eds.), Oxford University Press, 439-513.

D. Dubois, H. Prade (1987) Necessity measures and the resolution principle, IEEE Trans. Systems, Man and Cybernetics, Vol. 17, pp. 474-478.

D. Dubois, H. Prade (with the collaboration of H. Farreny, R. Martin-Clouaire, C. Testemale) (1988) Possibility Theory — An Approach to Computerized Processing of Uncertainty. Plenum Press, New York.

D. Dubois, and H. Prade (1990). The logical view of conditioning and its application to possibility and evidence theories. Int. J. of Approximate Reasoning, 4, 23-46.

D. Dubois, H. Prade (1991) Possibilistic logic, preferential models, non-monotonicity and related issues. Proc. of the Inter. Joint Conf. on Artificial Intelligence (IJCAI'91), Sydney, Aug. 24-30, 419-424.

D. Dubois, H. Prade (1995) Conditional objects, possibility theory and default rules'. In: Conditionals: From Philosophy to Computer Sciences (G. Crocco, L. Fariñas del Cerro and A. Herzig, eds.), Oxford University Press, 301-336.

D.M. Gabbay (1985) Theoretical foundations for non-monotonic reasoning in expert systems. In: Logics and models of Concurrent Systems (K.R. Apt, ed.), Springer Verlag, 439-457.

P. Gärdenfors (1988) Knowledge in Flux — Modeling the Dynamic of Epistemic States. MIT Press, Cambridge, MA.

P. Gärdenfors, D. Makinson (1994) Nonmonotonic inference based on expectations. Artificial Intelligence, 65, 197-245.

M. Goldszmidt, J. Pearl (1990) On the relation between rational closure and System Z. Proc. of the 3rd Inter. Workshop on Nonmonotonic Reasoning, South Lake Tahoe, 130-140.

E. Hisdal (1978) Conditional possibilities independence and noninteraction. Fuzzy Sets and Systems 1, 283-297.

S. Kraus, D. Lehmann, M. Magidor (1990) Nonmonotonic reasoning, preferential models and cumulative logics. Artificial Intelligence, 44, 167-207.

J. Lang (1991) Logique possibiliste: aspects formels, déduction automatique, et applications. PhD Thesis, Universite P. Sabatier, Toulouse, France, January 1991.

D. Lehmann (1989) What does a conditional knowledge base entail? Proc. of the 1st Inter. Conf. on Principles of Knowledge Representation and Reasoning (KR'89), Toronto, 357-367.

D. Lehmann, M. Magidor (1992) What does a conditional knowledge base entail? Artficial Intelligence, 55, 1-60.

D.L. Lewis (1973) Counterfactuals. Basil Blackwell Ltd, Oxford.

J. Lin and Mendelzon (1992). Merging databases under constraints. International Journal of Cooperative Information Systems. Vol. 7(1), 55-76.

D. Makinson (1989) General theory of cumulative inference. In: Non-Monotonic Reasoning in (M. Reinfrank, J. De Kleer, M.L. Ginsberg, E. Sandewall, eds.), LNAI 346, Springer Verlag, Berlin, 1-18.

S. Nahmias (1978) Fuzzy variables. Fuzzy Sets and Systems, 97-110.

J. Pearl (1990) System Z: A natural ordering of defaults with tractable applications to default reasoning. Proc. of the 3rd Conf. on Theoretical Aspects of Reasoning About Knowledge (TARK'90), San Mateo, CA, 121-135.

N. Rescher (1976) Plausible Reasoning: An Introduction to the Theory and Practice of Plausibilistic Inference. Van Gorcum, Amsterdam.

Y. Shoham (1988) Reasoning About Change — Time and Causation from the Standpoint of Artificial Intelligence. MIT Press, Cambridge, MA.

G.L.S. Shackle (1961) Decision, Order and Time in Human Affairs. Cambridge University Press, Cambridge, UK.

W. Spohn (1988) Ordinal conditional functions: A dynamic theory of epistemic states. In: Causation in Decision, Belief Change, and Statistics (W.L. Harper, B. Skyrms, eds.), Kluwer Academic Publ., Dordrecht, 105-134.

R.R. Yager (1992) On the specificity of a possibility distribution. Fuzzy Sets and Systems, 50, 279-292.

L.A. Zadeh (1978) Fuzzy sets as a basis for a theory of possibility. Fuzzy Sets and Systems, 1, 3-28.

ON THE SOLUTION OF FUZZY EQUATION SYSTEMS

H.-J. Lenz and R. Müller
Free University Berlin, Berlin, Germany

ABSTRACT

We consider a set of mixed linear and non-linear equations, which arise typically in
operative controlling. The variables in the single equations are connected by arithmetic
operations. In order to allow for imprecision, each variable is modelled as a fuzzy set
with a given membership function. Given a vector of observed variables and an equation
system with terms built up by fuzzy sets, a controlling decision is to be made, whether
the data set is 'consistent' with the equation system or not.

We first give a typical example where the functional relationships link various micro-
economic indicators together. Next we describe the nature of operative controlling and
give some formal definitions. The fuzzy set theory is presented limited to what is needed
for controlling. An algorithm is presented which has as input a real data set and a fuzzy
equation system, and computes consistent values for all variables, if a simultaneous
solution exists. Otherwise, it is flagged, that the data is inconsistent with the fuzzy
model. We close with various scenarios illustrating the (reasonable) behaviour of the
algorithm for solving such equations in the context of operative controlling.

1. ILLUSTRATING EXAMPLE

Consider the values of the following business indicators, which are printed together with (absolute) errors in the variables. It is reasonable to imagine, that different sources supplied the data set, which are considered as different sources of partial information. The errors reflect the imprecision with which each variable is metered or observed, judged by an 'overall system observer' or by the controller:

Data Set with error rates:
Expenditures $K = 40 \pm 10$,
Sales $U = 70 \pm 10$,
Profit $G = 10 \pm 5$,
Profit-turnover ratio $R = 0.1 \pm 0.07$.

We take a simplified variant of a business indicator model as our simultaneous equation system. Note that the two equations are given by definition.
Model:
(1) Profit = Sales - Expenditures,
(2) Profit-turnover ratio = Profit/ Sales

The reduced form of this structural equation system **G** can be written as follows and visualised as in the graph below:

$$\begin{bmatrix} G \\ R \end{bmatrix} = \mathbf{G}\left(\begin{bmatrix} K \\ U \end{bmatrix} \right)$$

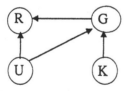

Fig. 1: Model Graph of a simplified business indicator model

Given a data set, a model and a set M of corresponding membership functions, the question arises, whether the data is 'consistent' with M according to a given metric or not. The solution of this problem will picked up later.

2. OPERATIVE CONTROLLING

The objective (target) of operative controlling is to detect 'irregularities' in the (operative) data produced by the various business processes. In a slightly more formal way one can define:

DEF. 1: Irregularities
Deviations between planned and real figures, which would be avoidable, if the business process where kept under control

There are many causes why irregularities can happen on the shop floor, for instance:

• Environment,
• Mismanagement,
• Illegal actions of single employees,
• Errors of any kind.

The controlling process itself is performed in 5 steps:
1. Based on a quantitative model, derive planned figures
2. Collect real data from the business processes of the last period
3. Compute a measure of deviation between planned and realised figures.
4. In case of 'significant' or 'large' deviations try to find out what happened by a causal analysis.
5. Adapt the control or decision processes governing the business processes.
We shall concentrate on steps 1 to 3.

3. MODELLING

Let us consider numerical data captured from the operative level of business processes. Examples are costs, prices, stocks, rates, balances etc. Such quantities are called 'business objects' or 'controlling objects', cf. Kluth[1996]. Instead of assuming that these variables have crisp values, they are represented as Fuzzy Sets. They reflect the imprecision of numerical data from the controllers point of view, and implicitly weight the 'reliability' of the various sources where the data come from.

Usually, the variables or 'controlling objects' are not isolated, but highly interconnected. The corresponding relationships are due to institutional or behavioural conditions or are simply given by definitions. For instance, the total salary of a firm is the sum of salaries of departmental sums, the advertising is about 12% of the annual costs and turnover is equal by definition to sales times price per unit. Modelling of such functional relationships is equivalent to operations on Fuzzy Sets. Fortunately enough, operative controlling is exclusively based on the set of arithmetic operations $= \{+,-,*,/\}$.

From a practical point of view there is no need for transformations like power(), log, exp, or √. Evidently, this simplifies the fuzzy modelling of the controlling process.
Next, we put together some theory of fuzzy sets, limited to what is really needed in our context, cf. Bandemer and Gottwald[1993].

DEF. 2: Fuzzy Set μ
A Fuzzy Set μ of reals is represented by the membership function $\mu: \mathbf{R} \rightarrow \mathbf{R}_{[0,1]}$.

DEF. 3: Fuzzy Number μ
A Fuzzy Set μ of reals is called Fuzzy Number, if
a) $\exists\, x \in \mathbf{R}$ with $\mu(x) = 1$ (normal Fuzzy Set) and
b) $\forall\, a,b,c \in \mathbf{R}$ with $b \geq c \geq a \Rightarrow \mu(c) \geq \min\{\mu(a), \mu(b)\}$ (convex level sets).

Normal Fuzzy Sets on $\mathbf{R}$ are important for specifying membership functions and are denoted by $\mathbf{F}(\mathbf{R})$ in the following. Next we need to define a sub-domain of a fuzzy set, where the membership function is positive, i.e. we want to characterise the 'relevant' range of a controlling object. In this way the controller fixes the span of imprecision.

DEF. 4: Support
Let $\eta \in \mathbf{F}(\mathbf{R})$. Then supp $(\eta)=\{x \in \mathbf{R}|\eta(x)>0\}$ is called support of the Fuzzy-Set η.

As a simple example take the information 'Consumption rate about 30 with an error of 20'.
This is modelled as a Fuzzy Set with supp $(\mu_1)=\{x \in \mathbf{R}|\mu_1(x)>0\}$ and a peak at 30, i.e. $\mu_1(30)=1$.

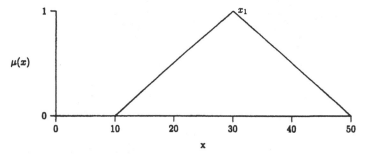

Fig. 2: Membership Function μ_1 of the fuzzy number $x_1=30\pm20$

Arithmetic operations on fuzzy sets are based on the famous extension principle of Zadeh[1975]. Consider for example aggregation over rates of consumption linked to departments. Let the total consumption metered be z and the department rates equal to x_1, x_2. Due to aggregation we have $z = x_1 + x_2$. If x_1, x_2 and z are modelled as fuzzy numbers with membership functions μ_1, μ_2 and μ_z, the algebraic equation $z = x_1 + x_2$ is corresponding to the equation $\mu_z \equiv \mu_1 \oplus \mu_2$. The operator '$\oplus$' combines the membership functions μ_1, μ_2 if summation is wanted and '$\equiv$' reflects the 'is equal to '- relation for fuzzy sets.

The basics of arithmetic operations are explained in detail in Bandemer, Gottwald [1989] and Kruse et al. [1993].

DEF. 5: Extension Principle

Let $k \in N$, μ_1, μ_2,...,μ_k be normal Fuzzy Sets and $\Phi : R^k \rightarrow R$. Then the Fuzzy Set

$$\Phi(\mu_1,...,\mu_k)(x) := \sup\{\min\{(\mu_1(x_1),...,\mu_k(x_k)\}|(x_1, x_2,...,x_k) \in R^k\}$$

with $\Phi(x_1, x_2,..., x_k) = x$ for all $x \in R$ is called range of $(\mu_1, \mu_2,...,\mu_k)$ given Φ.

If the principle of extension is applied on the aggregation of the consumption rates above, it follows:

$$\mu_+(x) := (\mu_1 \oplus \mu_2)(x) = \sup\{\min\{\mu_1(x_1),\ \mu_2(x_2)\}\ |\ x_1+x_2 = x,\ x_1,x_2 \in R\}.$$

In a similar way the other arithmetic operations can be performed, cf. Kruse et al. [1993].

We illustrate cross-sectional aggregation by summing the fuzzy consumption rates of two departments.

Example 1: Cross-sectional Aggregation over two departments

Let the metered consumption rates given as follows together with (absolute) errors:

$x_1 = 30 \pm 10$ and
$x_2 = 30 \pm 20$.

The variables x_1, x_2 are interpreted as fuzzy numbers with $supp_1 = 30 \pm 20$ and $supp_2 = 30 \pm 10$. Using the Extension Principle of Zadeh, it follows for $z = x_1 + x_2$ on the left hand side of this equation:

$\hat{z} = 60 \pm 30$. Note, that this fuzzy number can be considered as an 'estimate' or 'prediction of z derived from the evaluation by operations on fuzzy terms on the right hand side of $z = x_1 + x_2$. The predicted value $\hat{z}$ can be compared by the controller with the realisation $z \pm e_z$, where e_z is the prior error rate linked to the observation z and expressing the imprecision of this quantity.

The graphs of the three membership functions are given below.

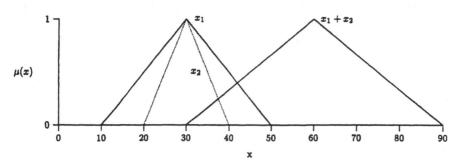

Fig. 3: Membership functions μ_1, μ_2 and $\mu_{\hat{z}}$ ($z = x_1 + x_2$)

We can now give a less vague description of the main objective of the prediction phase of operative controlling.

Given a data set $\mathbf{D} = (x_0, z_0)$, a system $G(x,z) = 0$ of linear or non-linear equations and a corresponding set $\mathbf{M}$ of fully specified membership functions $\eta \in \mathbf{F(R)}$, find $(\hat{x}, \hat{z})$ subject to $G(\hat{x}, \hat{z}) = 0$ if it exists. If such a pair $(\hat{x}, \hat{z})$ exists, we say, that $\mathbf{D}$ is consistent with respect to G given $\mathbf{M}$, otherwise inconsistent.

4. SINGLE EQUATION SYSTEMS

First we turn to a single equation. Consider for example $z = g(x_1, x_2, ... x_p)$. The main idea is to use a definition of inconsistency of two fuzzy sets. Roughly speaking, two fuzzy sets are inconsistent if they don't intersect or their distance is larger than a given upper bound. If we assume that g is a separable function, we can solve for each variable $x_1, x_2, ... x_p$, or z involved in g and compute the corresponding fuzzy set of the right hand side of each resulting equation. The intersection of the two fuzzy sets, one linked to the left hand side (LHS), the other to the right hand side (RHS) is determined and the corresponding membership function $\tilde{\mu}(x)$ computed. Renormalising of a fuzzy set x with membership function $\mu(x)$ is necessary, because it can happen that max $\tilde{\mu}(x) < 1$. Therefore the re-normalised $\hat{x}$ can be determined as follows:

if $x = \varnothing$ then $\hat{x} = \varnothing$, else $\mu_{\hat{x}} = \mu_x / \max \mu_x$. If the equation is non contradictive to the data ('consistent'), than the various intersections must be non-empty. Otherwise the equation has no solution, cf. Müller [1999]. As a convergence criterion we use the condition 'if $\hat{z} \subseteq g(\hat{x}_1, \hat{x}_2, ..., \hat{x}_p)$ and $\hat{x}_i \subseteq g_i(\hat{x}_i, ..., \hat{x}_{i-1}, \hat{x}_{i+1}, ..., \hat{x}_p, \hat{z})$' then convergence () is true.

Finding $(\hat{x}_1, \hat{x}_2, ..., \hat{x}_p, \hat{z})$ or checking consistency of $(x_1, x_2, ... x_p, z)$ with respect to g given fully specified membership functions $\mu_1, \mu_2, ..., \mu_p, \mu_z$ is performed using the following algorithm FES.

Algorithm FES

(* Find a solution of a non-linear equation of fully specified fuzzy sets $x_1, x_2, ... x_p, z$ or set flag

'Inconsistency' true.

A distance measure $\delta()$ between a pair of fuzzy sets is given.

convergence criterion converge( ) *)

input: $(x_1, x_2, ... x_p, z)$

output: $(\hat{x}_1, \hat{x}_2, ..., \hat{x}_p, \hat{z})$

external $\delta(\ )$

begin

      loop

         $\tilde{z} := g(x_1, ..., x_p)$ with membership function $\mu_{\tilde{z}}$

         if $\delta(z, \tilde{z}) \le \varepsilon$ then $z^c := z \cap \tilde{z}$ else stop('Inconsistency')

         for i:=1 to p

             $\tilde{x}_i := g_i(x_1, ... x_{i-1}, x_{i+1}, ... x_p, z)$ with membership function $\mu_{\tilde{x}i}$

             if $\delta(x_i, \tilde{x}_i) \le \varepsilon$ then $x_i^c := x_i \cap \tilde{x}_i$ else stop('Inconsistency')

         endfor

         $\hat{z} := \text{renorm}(z^c)$

         $\hat{x}_i := \text{renorm}(x_i^c)$ for all i:=1,...,p

         exitloop if $\text{converge}((\hat{z}, \hat{x}_i)_{i=1,...p})$

      endloop

end

This algorithm has some appealing characteristics, which is analogue to an algorithm which is based on random variables, cf. Lenz, Rödel [1999].

1. The supports monotonically contract, i.e. $\text{supp}(x_i) \supseteq \text{supp}(\hat{x}_i)$ for all i=1,2,...,p and $\text{supp}(z) \supseteq \text{supp}(\hat{z})$.

2. Membership functions are shifted into the 'right' direction and this is done coherently. For example, if two factors are 'too small' and/or a product 'too large', then the factors are increased while the product is decreased. A 'zigzagging' of de- and increase of membership functions is excluded.

3. The size of a shift of a membership function is (roughly) proportional to its support. Especially, crisp numbers are left unchanged.

4. Consistent data is left unchanged by the algorithm (invariance property).

5. The algorithm is associative, i.e. the sequence of variables in the inner loop does not affect the solution, if it exists.

We present now some examples to illustrate the behaviour of algorithm FES for a simple (linear) equation with three variables. For U=G+K we consider the following scenarios:

S1: 'Standard inconsistency in measured data with 'normal' errors in the variables'
    (note that 15+35=50<65)

original data:	corrected data:
G=15±5	$\hat{G}$ = 18.6 -3.6/+1.4
K=35±5	$\hat{K}$ = 38.6 - 3.6/+1.4
U=65±10	$\hat{U}$ = 57.5 ± 2.5

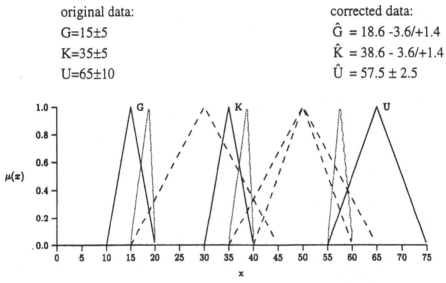

Fig. 4: Standard inconsistency in measured data with 'normal' errors in the variables. Membership functions corresponding to original data (unbroken line), corrected data (dotted line), and right hand side (dashed line)

S2: 'Reduced inconsistency in measured data with 'normal' errors in the variables'
    (note that 15+35<55)

original data:	corrected data:
G=15±5	$\hat{G}$ = 16.1 -6.1/+3.9
K=35±5	$\hat{K}$ = 36.1 - 6.1/+3.9
U=55±10	$\hat{U}$ = 52.5 ± 7.5

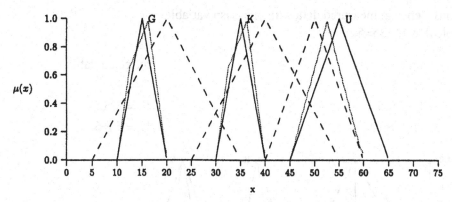

Fig. 5: Reduced inconsistency in measured data with 'normal' errors in the variables. Membership functions corresponding to original data (unbroken line), corrected data (dotted line), and right hand side (dashed line)

S3: 'Consistency in measured data with 'normal' errors in the variables' (note that 15+35=50)

original data: corrected data:

G=15±5 $\hat{G} = 15±5$

K=35±5 $\hat{K} = 35±5$

U=50±10 $\hat{U} = 50±10$

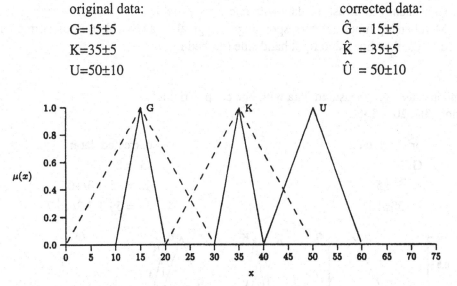

Fig. 6: 'Consistency in measured data with 'normal' errors in the variables' Membership functions corresponding to original data (unbroken line), corrected data (dotted line), and right hand side (dashed line)

S4: 'Consistency in measured data with one crisp variable'
(note that 15+35=50)

original data:	corrected data:
G=15	$\hat{G}$ = 15
K=35±5	$\hat{K}$ = 35±5
U=50±10	$\hat{U}$ = 50±5

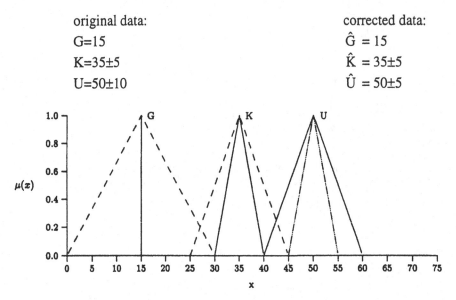

Fig. 7: Consistency in measured data with one crisp variable.
Membership functions corresponding to original data (unbroken line), corrected
data (broken line), and right hand side (dashed line)

S5: 'Inconsistency in measured data with one crisp variable'
(note that 20+35>50)

original data:	corrected data:
G=20	$\hat{G}$ = 20
K=35±5	$\hat{K}$ = 33-3.3/+6.7
U=50±10	$\hat{U}$ = 53.3-3.3/+6.7

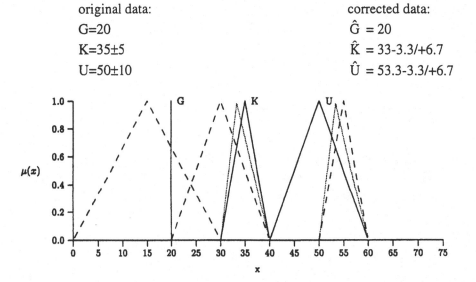

Fig. 8: Inconsistency in measured data with one crisp variable.
Membership functions corresponding to original data (unbroken line), corrected data (broken line), and right hand side (dashed line)

S6: 'Inconsistency in measured data and small supports
(note that 15+35>55)

original data: corrected data:

G=15±1 $\hat{G}$ = 15-1.5/+0.5

K=35±2 $\hat{K}$ = 36.2-2.2/+0.8

U=55±5 $\hat{U}$ = 51.8-1.8/+1.2

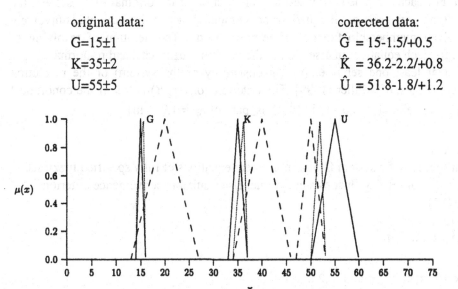

Fig. 9: Inconsistency in measured data and small supports .
Membership functions corresponding to original data (unbroken line), corrected data (broken line), and right hand side (dashed line)

5. SYSTEMS OF NON-LINEAR EQUATIONS

We assume to have a system G of m equations with a total of p variables (fuzzy sets); we don't distinguish anymore between variables on the left and right side of an equation and use the uniform notation $x_1, x_2, ... x_p$. Each equation g of G is assumed to be a separable function. Therefore we can solve uniquely for each variable $x_1, x_2, ... x_p$ involved in any g and compute the corresponding fuzzy set of the right hand side of each equation. For each variable (fuzzy set) the intersection of the original fuzzy set and the corresponding fuzzy sets of all equations are computed, where this variable is involved. This resolved membership function is re-normalised. The iteration of resolving of variables for each equation, intersection of the resulting fuzzy sets, and re-normalising is stopped if at least one set is empty (inconsistency of the system) or the resolution process converges, cf. Müller [1999]. The criterion converge() is true if the condition '$\hat{x}_i^k \subseteq RHS_{ij}^k$ for all i=1,2,...,p and $j \in \{j| \hat{x}_i^k$ is_part_of g_j, j=1,2,...,m}.

Algorithm FSS
(* Find a solution of a system G of m non-linear equations of fully specified fuzzy sets x_i, i=1,2,...,p   or set flag 'Inconsistency' true. Use a suitable convergence criterion converge() *)
input: $(x_1, x_2, ... x_p)$
output: $(\hat{x}_1, \hat{x}_2, ..., \hat{x}_p)$
begin
         k:=0; $\hat{x}_i^k := x_i$ for all i=1,2,...,p
    loop
         k:=k+1
         for j=1,2,...,m
                 solve equation g_j w.r.t. $\hat{x}_i^{k-1}$ if $\hat{x}_i^{k-1}$ is_part_of g_j
                 compute $\tilde{x}_{ij} := RHS_{ij}^{k-1}$ by fuzzy arithmetic
        endfor
        for i=1,2,...,p
                 compute $x_i^c := (\bigcap_{j \in G'} \tilde{x}_{ij}) \cap \hat{x}_i^{k-1}$, where $G'=\{g_j \in G| \tilde{x}_{ij}$ is_part_of $g_j\}$
                 $\hat{x}_i^k := renorm(x_i^c)$
        endfor
        exitloop if $((\exists i: \hat{x}_i^k = \emptyset)$ or converge$((\hat{x}_i^k)_{i=1,...p})$
    endloop
end

We close with some illustrations how the algorithm operates on a simple equation system.

We use the following business indicator model:

$$G = U-K$$
$$R = G/U.$$

The data and the error rates are given as follows:

original data:

$G = 10\pm5$

$K = 40\pm10$

$U = 70\pm10$

$R = 0.1\pm0.07$

corrected data:

$\hat{G} = 10.15-0.15/+0.09$

$\hat{K} = 48.86-0.1/+014$

$\hat{U} = 60.04-0.04/+0.2$

$\hat{R} = 0.169-0.003/+0.001$

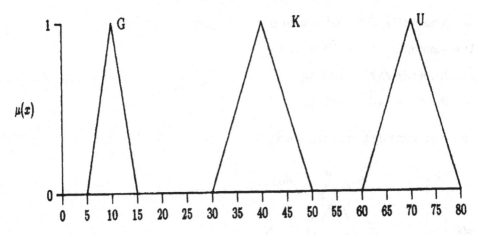

Fig. 10: Membership functions of original data G, K, U

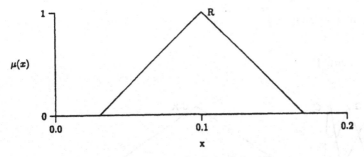

Fig. 11: Membership function of original data R

In the first pass the algorithm takes equation 1 and solves it for each variable, i.e.

$\tilde{G}^{(1)} := U - K = 30 \pm 20$,

$\tilde{U}^{(1)} := G + K = 50 \pm 15$,

$\tilde{K}^{(1)} := U - G = 60 \pm 15$.

Then equation two is processed:

$\tilde{R}^{(1)} := G / U = 0.14\text{-}0.08/\text{+}0.1$

$\tilde{U}^{(2)} := G / R = 100\text{-}66.6/\text{+}200$,

$\tilde{G}^{(2)} := R * U = 7\text{-}4/\text{+}5$.

Note that a superscript reflects the number of re-computations of a specific variable. Then the membership functions of the 4 fuzzy sets are re-normalised:

$\hat{G} := \operatorname{renorm}(G \cap \tilde{G}^{(1)} \cap \tilde{G}^{(2)})$ with $\mu_{\hat{G}}$

$\hat{U} := \operatorname{renorm}(U \cap \tilde{U}^{(1)} \cap \tilde{U}^{(2)})$ with $\mu_{\hat{U}}$

$\hat{K} := \operatorname{renorm}(K \cap \tilde{K}^{(1)})$ with $\mu_{\hat{K}}$

$\hat{R} := \operatorname{renorm}(R \cap \tilde{R}^{(1)})$ with $\mu_{\hat{R}}$

The corresponding graphs are given below.

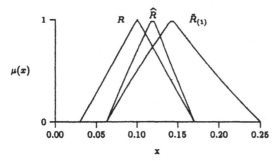

Fig. 12: R, $\tilde{R}^{(1)}$ (=G/U) and $\hat{R}$

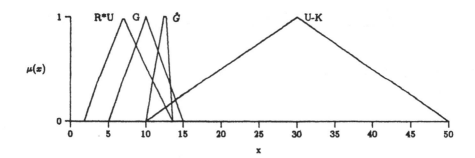

Fig. 13: G, $\tilde{G}^{(1)}$ (=R*U), $\tilde{G}^{(2)}$ (=U-K) and $\hat{G}$

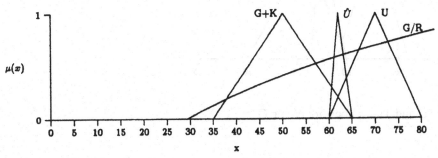

Fig. 14: U, $\tilde{U}^{(1)}$ (=G+K), $\tilde{U}^{(2)}$ (=G/R) and $\hat{U}$

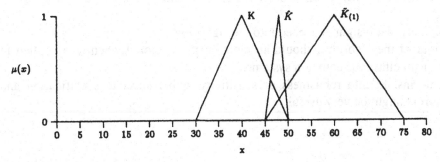

Fig. 15: K, $\tilde{K}^{(1)}$ (=U-G), and $\hat{K}$

At least one further pass is necessary, because converge($\hat{G}^k$) is not fulfilled as can be seen from Fig. 17. This is due to the effect, that the fuzzy sets $\hat{R}*\hat{U}$ and $\hat{U}-\hat{K}$ does not cover $\hat{G}$, which is a sufficient condition for convergence. The statement ' A covers B ' is defined by $\mu_B(x) \leq \mu_A(x)$ for all $x \in \mathbf{R}$.

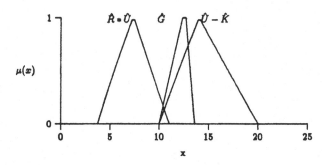

Fig. 16: $\hat{R} * \hat{U}$ and $\hat{U} - \hat{K}$ does not cover $\hat{G}$

6. PERSPECTIVES

There some points of interest for further studies. First of all, appropriate distance measures beyond Kruse-Meyer distance or Haussdorff metric between fuzzy sets should be analysed.

Some more analytical insight is needed into the algorithm.

The behaviour of the algorithm should be more deeply explored when it is applied to recursive and simultaneous equation systems.

Finally, some insight into robustness or sensitivity with respect to specification and numerical errors might be very useful.

7. REFERENCES

1. M. Kluth (1996) Wissensbasiertes Controlling von Fertigungseinzelkosten, Gabler, Wiesbaden

2. H. Bandemer and S. Gottwald (1993) Einführung in Fuzzy-Methoden, Akademie Verlag,Berlin

3. R. Müller (1999) Controlling, Planung und Prognose mit unscharfen Daten, MSc Thesis, Institut für Produktion, Wirtschaftsinformatik und Operations Research, Freie Universität Berlin, Berlin

4. R. Kruse, J. Gebhardt and F. Klawonn (1993) Fuzzy-Systeme, Teubner, Stuttgart

5. H.-J. Lenz and E. Rödel (1999) Controlling based on Stochastic Models, in this volume

6. L.A. Zadeh (1965) Fuzzy Sets. Information and Control, 8, 338-353

LEARNING FUZZY MODELS AND POTENTIAL OUTLIERS

M.R. Berthold
University of California at Berkeley, Berkeley, CA, USA

Abstract. Outliers or distorted attributes very often severely interfere with data analysis algorithms that try to extract few meaningful rules. Most methods to deal with outliers try to completely ignore them. This can be potentially harmful since the very outlier that was ignored might have described a rare but still extremely interesting phenomena. In this paper we describe an approach that tries to build an interpretable model while still maintaining all the information in the data. This is achieved through a two stage process. A first phase builds an outlier-model for data points of low relevance, followed by a second stage which uses this model as filter and generates a simpler model, describing only examples with higher relevance, thus representing a more general concept. The outlier-model on the other hand may point out potential areas of interest to the user. Preliminary experiments indicate that the two models in fact have lower complexity and sometimes even offer superior performance.

1 Introduction

Many datasets obtained from real-world systems contain distorted elements, for example due to errors in measurements, sensor-failures, or simple recording problems. If the resulting data is to be analyzed by means of extracting an interpretable model from this data, these so-called "outliers" are often difficult to ignore. Many existing methodologies to build models from data will try to incorporate the outliers, making the resulting model more complex and harder to interpret [4,6,10,11,13].

In this paper an approach is discussed which aims to model an evolving set of potential outliers through an additional outlier-model. This results in two models, one representing a more general concept which offers better understandability, and the other concentrating on potential outliers or regions with low evidence in the observed data. The presented approach is based on sets of fuzzy rules as a way to model imprecise relationships [14,15]. Based on an existing, fast algorithm that constructs fuzzy rule sets from data [7], the presented

* M. Berthold was in part supported by a stipend of the "Gemeinsame Hochschulsonderprogramm III von Bund und Ländern" through the DAAD, BISC, and the DFG.

approach builds a model for the entire data in a first phase. After completion this model is analyzed automatically and the parts with low relevance in respect to the training data are moved to the outlier model. The outlier model is then used as a filter for the second phase. The second phase generates a more general model, representing the data points with higher evidence. The final pair of models consists of one model for potential outliers and one model which represents the more general behavior. The resulting general model is less complex and therefore easier to interpret whereas the outlier-model may point out potential areas of interest to the user.

We will start out by describing a method to build fuzzy models based on example data, followed by a description of the two-stage fuzzy model generation. Thereafter we present results on two datasets, demonstrating the effects of the proposed methodology.

2   Learning Fuzzy Models from Data

In the following we will concentrate on n-dimensional feature spaces with one dependent variable which is either continuous (for an approximation problem) or denotes a set of classes (in case of a classification task). The goal of the discussed methods is to generate a set of rules which describe some available example data. The used rules describe an implication:

$$\mathcal{R} : \textbf{if } antecedent \textbf{ then } consequent \quad \textbf{with } weight \ w$$

and we will focus on the most commonly used form of rules, consisting of an antecedent in form of a constraint on the input variables:

$$x_1 \textbf{ is } A_1 \textbf{ and } \cdots \textbf{ and } x_n \textbf{ is } A_n \qquad \text{or short}: \quad \boldsymbol{x} \textbf{ is } \boldsymbol{A}$$

where the A_i $(1 \leq i \leq n)$ describe fuzzy sets which are defined through their membership functions $\mu_{A_i} : \mathbb{R} \to [0,1]$. The consequent of such rules assigns a certain class c in case of a classifier

$$\mathcal{R} : \textbf{if } \boldsymbol{x} \textbf{ is } \boldsymbol{A} \textbf{ then class } c$$

where c denotes one of C possible classes. In case of a fuzzy graph describing a functional relationship, the consequent specifies a granule for the dependent or output variable

$$\mathcal{R} : \textbf{if } \boldsymbol{x} \textbf{ is } \boldsymbol{A} \textbf{ then } y \textbf{ is } B$$

where B again denotes a fuzzy set, defined through a membership function $\mu_B :$ $\mathbb{R} \to [0,1]$. Of interest is sometimes also a weight parameter w which usually represents the percentage of training patterns explained by this rule. In the following two sections the differences between fuzzy models used for classification and approximation are highlighted before an algorithm to construct both types of models is discussed.

2.1 Fuzzy Classifiers

For a classifier the consequent assigns class c out of a finite number of classes C. The degree of membership for a certain pattern x for rule $\mathcal{R}$ is simply computed through the membership degree of the antecedent where the conjunction $\wedge$ is often implemented through the minimum:

$$\mu_{\mathcal{R}}(x) = \mu_{A_1}(x_1) \wedge \cdots \wedge \mu_{A_n}(x_n) = \min\{\mu_{A_1}(x_1), \cdots, \mu_{A_n}(x_n)\}$$

In this n-dimensional feature space the area of influence of each rule $\mathcal{R}$ is thus specified by the vector of membership functions $(A_1, \cdots, A_n)$. A set of rules is then defined through

$$\mathbf{R} = \{\mathcal{R}^{(r)} \mid 1 \leq r \leq R\}$$

where R indicates the number of rules. Of course, each of these rules only assigns one specific class $c^{(r)}$ out of $\{1, \cdots, C\}$.

The combined degree of membership for a certain class c is computed through a disjunction $\vee$ which is implemented through the maximum:

$$\mu_{\mathbf{R}}^c(x) = \bigvee\{\mu_{\mathcal{R}^{(r)}}(x) \mid c^{(r)} = c\} = \max\{\mu_{\mathcal{R}^{(r)}}(x_1) \mid c^{(r)} = c\}$$

In a prediction scenario with equal risk, the class with the highest degree of membership would typically be chosen.

2.2 Fuzzy Approximators

For approximation the rule "if x is A then y is B" can also be seen as a constraint on a joint variable (x, y), that is,

$$(x, y) \text{ is } A \times B$$

where $\times$ denotes the Cartesian product. The membership function of $A \times B$ is again given using $\wedge$ as a conjunction operator:

$$\mu_{A \times B}(x, y) = \mu_A(x) \wedge \mu_B(y) = \min\{\mu_A(x), \mu_B(y)\}$$

or, more precisely

$$\mu_{A \times B}(x, y) = \min\{\mu_{A_1}(x_1), \cdots, \mu_{A_n}(x_n), \mu_B(y)\}$$

Zadeh [14] also uses the term *fuzzy point* to denote such a rule $A \times B$. A collection $\mathbf{R}$ of rules can now be regarded as forming a superposition of R fuzzy points:

$$(x, y) \text{ is } (A_1 \times B_1 + \cdots + A_R \times B_R)$$

where $+$ denotes the disjunction operator (usually defined as maximum). Zadeh calls this characterization of a dependency *fuzzy graph*, because the collection of

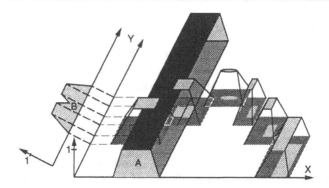

Fig. 1. Interpolation as the intersection of a fuzzy graph with a cylindrical extension of A.

rules can be seen as a coarse representation of a functional dependency f^* of y on x. This fuzzy graph f^* can thus be defined as:

$$f^* = \sum_{r=1}^{R} (A_r \times B_r)$$

Note that individual rules do not depend on a common set of membership functions on the input variables. Instead each fuzzy point is described through an individual set of membership functions A_r which, especially in high dimensional cases, restrict only few attributes. This makes fuzzy graphs a very compact description, especially compared to the approaches that use a "global grid", defined through only one set of membership functions for each input variable (see for example [12,6]).

The task of interpolation, that is, deriving a linguistic value B' for y given an arbitrary linguistic value A for x and a fuzzy graph f^*:

$$\frac{x \text{ is } A}{f^* \text{ is } \sum_{j=1}^{m} A_j \times B_j}$$
$$y \text{ is } B'$$

results in an intersection of the fuzzy graph f^* with a cylindrical extension of the input fuzzy set A. Figure 1 shows an example. This functional dependency can be computed through:

$$\mu_{B'} = \mu_{f^*(A)}(y) = \sup_{\infty}\{\mu_{f^*}(x,y) \wedge \mu_A(x)\}$$
$$= \sup_{\infty}\{(\mu_{A_1 \times B_1}(x,y) \vee \cdots \vee \mu_{A_r \times B_r}(x,y)) \wedge \mu_A(x)\}$$

If a crisp output is desired the resulting fuzzy set B' has to be defuzzified, a common approach here is the "Center of Gravity" method which simply computes to:

$$\tilde{y} = \int_y y \cdot \mu_{B'}(y) dy$$

In the following a strategy to construct such fuzzy rule sets based on example data is described in detail.

2.3 Learning Fuzzy Rules

In both cases, that is classification or approximation, the resulting training data consists of an input vector x, together with a vector of membership values μ^{target} for the desired classes or output granules. The algorithm described here also allows for some – but not necessarily all – features having predefined granulation and is based on two methods to build fuzzy rules and fuzzy graphs [7,2]. As a-priori information we therefore have the following:
for the output:

- either the number C^{out} of classes in case of a classifier, or in case of a continuous output variable its granulation into a set of C^{out} linguistic values, described through membership functions $\mu_c^{\text{out}} : \mathbb{R} \to [0,1]$, with $1 \leq c \leq C^{\text{out}}$.

  This also has implications for the possible consequents of the generated rules. In both cases, there is only a limited choices of possible consequents, either one out of C^{out} classes or one of C^{out} granules. In the following we will therefore universally denote a consequent by $g \in \{1, \cdots, C^{\text{out}}, \}$ possibly accompanied by a superscript indicating the rule's index.

and for the inputs:

- a number n_{granul} of features x_i ($1 \leq i \leq n_{\text{granul}}$) with a predefined granulation, defined through a set of C_i^{in} linguistic values, described through membership functions $\mu_i^{\text{in},c} : \mathbb{R} \to [0,1]$, with $1 \leq c \leq C_i^{\text{in}}$ and $1 \leq i \leq n_{\text{granul}}$.
- and a number n_{free} of features x_i ($n_{\text{granul}} + 1 \leq i \leq n = n_{\text{granul}} + n_{\text{free}}$) without a-priori defined granulation[1].

Example 1. In case of a three-dimensional feature space ($n = 3$) with a continuous output variable "pressure", one could define $C^{\text{out}} = 4$ granules, where the linguistic values "very low", "low", "medium", and "high" are defined through the membership functions μ_c^{out} ($c = 1, \ldots, 4$). The three input features could consist of one granulated feature ($n_{\text{granul}} = 1$) temperature, with three linguistic values "cold", "warm", and "hot" ($C_1^{\text{in}} = 3$) and another two free features x_2 and x_3 ($n_{\text{free}} = 2$) for two sensor values.

---

[1] For clarity of representation we assume that the features are ordered with respect to being granulated or not. This does, however, not affect generality of the discussed approach.

In addition to this information a set of training examples exists:

$$\mathbf{T} = \left\{ \left(\boldsymbol{x}^{(m)}, \boldsymbol{\mu}^{\text{target},(m)} \right) \mid 1 \leq m \leq M \right\}$$

where M indicates the number of examples in the set $\mathbf{T}$. In case of a classifier the values $\mu_c^{\text{target},(m)}$ indicate the degree of membership to class c for training example m. For a continuous output $\mu_c^{\text{target},(m)}$ specifies the desired degree of membership to granule c ($1 \leq c \leq C^{\text{out}}$).

The goal of the algorithm is thus to generate a set of rules $\mathbf{R}$, which describe the training data $\mathbf{T}$.

$$\mathbf{R} = \left\{ \mathcal{R}^{(r)} \mid 1 \leq r \leq R \right\}$$

Each of the R rules is specified through a set of constraints on the input domain and the index of the associated granule or class $g^{(r)}$ ($1 \leq g^{(r)} \leq C^{\text{out}}$):

$$\mathcal{R}^{(r)} : \text{if} \bigwedge_{i=1}^{n_{\text{granul}}} \text{cond}_{\text{granul},i}^{(r)} \text{ and } \bigwedge_{i=n_{\text{granul}}+1}^{n} \text{cond}_{\text{free},i}^{(r)} \text{ then } g^{(r)} \text{ with } w^{(r)}$$

where the constraint consists of two parts, $\text{cond}_{\text{granul},i}^{(r)}$ and $\text{cond}_{\text{free},i}^{(r)}$.

- The first part describes the restrictions on the granulated part of the feature space:

$$\text{cond}_{\text{granul},i}^{(r)} = x_i \text{ is } \bigvee_{c=1}^{C_i^{\text{in}}} \left(s_i^c \wedge \mu_i^{\text{in},c} \right)$$

and s_i^c specifies the applicable linguistic values for that specific feature[2]. If $s_i^c = 1$, value c is part of the condition, whereas $s_i^c = 0$ indicates its absence in $\text{cond}_{\text{granul},i}^{(r)}$.

Example 2. Assuming a linguistic variable "temperature" (index i) for rule r, the condition $\text{cond}_{\text{granul},i}^{(r)}$ could for example contain two linguistic values "cold" and "warm" out of the set of three possible values "cold" ($\mu_i^{\text{in},1}$), "warm" ($\mu_i^{\text{in},2}$), and "hot" ($\mu_i^{\text{in},3}$). Then $C_i^{\text{in}} = 3$ and $s_i^1 = 1$, $s_i^2 = 1$, and $s_i^3 = 0$. Rule $\mathcal{R}^{(r)}$ would thus have the following constraint (among others): $\mathcal{R}^{(r)} : \text{if } \cdots \text{temperature is (cold or warm)} \cdots \text{ then } \cdots$

In most cases $\text{cond}_{\text{granul},i}^{(r)}$ will actually contain all linguistic values ($s_i^\star = 1$), indicating that this feature is not restricted at all.

- The second part of the constraints $\text{cond}_{\text{free},i}^{(r)}$, are specified through:

$$\text{cond}_{\text{free},i}^{(r)} = \begin{cases} \text{true} \\ x_i \text{ is } < a_i^{(r)}, b_i^{(r)}, c_i^{(r)}, d_i^{(r)} > \end{cases}$$

which either specify no constraint at all, or a trapezoidal membership function in case a constraint was derived by the training algorithm.

---

[2] We will skip the rule's index (r) for sake of readability, it will be clearly defined from the context.

The algorithm to derive such a set of rules $\mathbf{R}$ from a set of examples $\mathbf{T}$ is described in the following section.

2.4 Algorithmic Details

The training algorithm operates sequentially, considering one example pattern after the other. At start no rules are existent, during training new rules will be introduced and existing rules will be fine tuned. In order to be able to guarantee termination of the algorithm, each rule keeps track of the pattern that originally triggered its creation. This so-called "anchor" is denoted $\lambda^{(r)}$ for rule $\mathcal{R}^{(r)}$. In addition we will use the following abbreviation:

$$\text{vol}(\mathcal{R}) := \int \mu_{\mathcal{R}}(x)\,dx$$

$$= \int \cdots \int \prod_{i=1}^{n_{\text{granul}}} \left(\bigvee_{c=1}^{C_i^{\text{in},c}} s_i^c \cdot \mu_{\text{granul},i}^{\text{in},c}(x_i) \right) \cdot \prod_{i=n_{\text{granul}}+1}^{n} \left(\mu_{\text{free},i}^{\text{in},c}(x_i) \right) \, dx_1 \cdots dx_n$$

which represents the volume rule $\mathcal{R}$ covers in the feature space,

$$\mu_{\mathcal{R}}(x) := \bigwedge_{i=1}^{n_{\text{granul}}} \left(\bigvee_{c=1}^{C_i^{\text{in},c}} s_i^c \wedge \mu_{\text{granul},i}^{\text{in},c}(x_i) \right) \wedge \bigwedge_{i=n_{\text{granul}}+1}^{n} \mu_{\text{free},i}^{\text{in},c}(x_i)$$

which represents the degree of membership of input x to the antecedent of rule $\mathcal{R}$, that is its degree of fulfillment, and finally

$$\text{maxindex}(\mu) := \arg\max\left\{ \mu_c \,|\, 1 \le c \le C^{\text{out}} \right\}$$

which determines the index of the component in μ with the maximum degree of membership.

During training for each example pattern $(x, \mu^{\text{target}}) \in \mathbf{T}$ two different scenario can now occur:

– COVERED: in this case a rule index r_{win} exists which already describes the new pattern, and also has the largest degree of membership among all rules:

$$\exists r_{\text{win}} : 1 \le r_{\text{win}} \le R : g^{(r_{\text{win}})} = \text{maxindex}(\mu^{\text{target}}) \wedge \mu_{\mathcal{R}^{r_{\text{win}}}}(x) > 0.0$$

$$\wedge\, \forall r' \ne r_{\text{win}} : g^{(r')} = \text{maxindex}(\mu^{\text{target}}) \rightarrow (\mu_{\mathcal{R}^{r_{\text{win}}}}(x) \ge \mu_{\mathcal{R}^{r'}}(x))$$

The core region of this rule will be increased along the non-granulated features to make sure it covers the new pattern (non-restricted free features do not need to be modified since they cover the pattern anyway):

$$\forall i : n_{\text{granul}} + 1 \le i \le n :$$

$$\text{cond}_{\text{free},i}^{r_{\text{win}}} \ne \text{true} \Rightarrow b_i^{\prime r_{\text{win}}} := \min\{b_i^{r_{\text{win}}}, x_i\} \wedge c_i^{\prime r_{\text{win}}} := \max\{c_i^{r_{\text{win}}}, x_i\}$$

and the weight of the rule will be incremented, thus keeping track of the number of patterns this rule explains: $w^{\prime r_{\text{win}}} := w^{r_{\text{win}}} + 1$.

COMMIT: in contrast to the above case no rule exists which describes the new pattern. In this case a new rule will be introduced (committed), describing the new example:

$$\mathcal{R}^{(R+1)} : \text{if true then } g^{(R+1)} := \text{maxindex}(\mu^{\text{target}}) \text{ with } w^{(R+1)} := 1$$

Obviously this rule is too general and will need to be specialized subsequently, through adaptation of the antecedent. (Note the weight $w^{(R+1)}$ which is initialized to 1, indicating that this rule explains one example pattern so far.) We also need to remember the pattern which triggered creation of this rule:

$$\lambda^{(R+1)} := x.$$

For simplicity we also remember the index of the linguistic values of all granulated features which cover the initial pattern x best:

$$\forall i : 1 \leq i \leq n_{\text{granul}} : c_{\lambda,i}^{(R+1)} := \arg \max \left\{ \mu_i^{\text{in},c}(x) \,|\, 1 \leq c \leq C_i^{\text{in}} \right\}$$

In addition to the modification and creation of rules describing patterns it is also necessary to actively avoid conflicts, that is, rules describing wrong relations need to be modified. This is done through a third step, called SHRINK, which modifies rules that incorrectly cover a pattern. More precisely, all rules $\mathcal{R}^{(r)}$ will be investigated if they fulfill

$$1 \leq r \leq R \wedge \mu_{g(r)}^{\text{target}} = 0 \rightarrow \mu_{\mathcal{R}^{(r)}}(x) = 0$$

that is, if the rule's consequent has a degree of membership equal to zero for this particular target μ^{target} then it also generates a degree of membership equal to zero for the corresponding input vector x. For each rule which violates this condition the following SHRINK-procedure will be performed:

- SHRINK: The goal is to specialize the conflicting rule $\mathcal{R}^{(r)}$ in a manner which inhibits the conflict. This is done through specialization of the rule, that is, the area of influence will be shrunk to exclude the new example pattern. To achieve this several alternatives exist, which are applied in the following order:
  - **S1** : Restrict an already constrained granulated feature. In this scenario it is assumed that the conflict can be resolved by tightening an already existing constraint of a granulated feature. Hence we restrict ourselves to:

$$\forall i_{S1} : 1 \leq i_{S1} \leq n_{\text{granul}}$$
$$\wedge \exists c : 1 \leq c \leq C_{i_{S1}}^{\text{in}} \wedge s_{i_{S1}}^c = 0$$
$$\wedge \exists c : 1 \leq c \leq C_{i_{S1}}^{\text{in}} \wedge s_{i_{S1}}^c = 1 \wedge \mu_{i_{S1}}^{\text{in},c}(x) = 0 \wedge c \neq c_{\lambda,i_{S1}}$$

that is, all granulated features for which at least one linguistic value is excluded from $\text{cond}_{\text{granul},i_{S1}}^{(r)}$ and at least one of the linguistic values in $\text{cond}_{\text{granul},i_{S1}}^{(r)}$ results in a degree of membership equal to zero but this linguistic value is not the one which covered the anchor of this rule during the initial commit. The last two requirements make sure that we can restrict this feature without completely erasing rule $\mathcal{R}^{(r)}$ or loosing coverage of its anchor. If no such feature can be found, alternative S2 is probed.

Otherwise we compute the loss in rule r's volume, assuming that feature i_{S1} is used to eliminate the conflict, resulting in a revised rule $\mathcal{R'}_{i_{S1}}^{(r)}$ which is similar to rule $\mathcal{R}^{(r)}$, the only difference being the constraint on the granulated feature i_{S1}:

$$\text{cond}_{\text{granul},i_{S1}}^{(r)} := x_{i_{S1}} \text{ is } \bigvee_{c=1}^{C_{i_{S1}}^{\text{in}}} \underbrace{\left(s_{i_{S1}}^c \wedge (\mu_{i_{S1}}^{\text{in},c}(x) = 0)\right)}_{=:s_{i_{S1}}^{\prime c}} \mu_{i_{S1}}^{\text{in},c}$$

all other constraints remain unchanged. From there we can compute the respective loss in rule r's volume:

$$\text{loss}_{i_{S1}} = \text{vol}\left(\mathcal{R}^{(r)}\right) - \text{vol}\left(\mathcal{R'}_{i_{S1}}^{(r)}\right)$$

We then choose the one feature $i_{S1,\text{best}}$ which minimizes this loss in volume, thus keeping the restricted rule as large as possible:

$$i_{S1,\text{best}} = \arg\min\{\text{loss}_{i_{S1}}\}$$

and replace rule $\mathcal{R}^{(\nabla)}$ in the new set of rules:

$$\mathbf{R}' := \mathbf{R} \cup \{\mathcal{R'}_{i_{S1,\text{best}}}^{(r)}\} \backslash \{\mathcal{R}^{(r)}\}.$$

Example 3. A rule $\mathcal{R}$ describes the fact that one buys oil when the temperature is either "cold" or "warm":

$$\mathcal{R} : \textbf{if } temperature \textbf{ is } cold \textbf{ or } warm \textbf{ then } buy\ oil$$

This rule might result in a conflict because an example household did not buy oil during "warm" temperatures. The revised rule would then be (through further restriction of feature *temperature*):

$$\mathcal{R}' : \textbf{if } temperature \textbf{ is } cold \textbf{ then } buy\ oil$$

S2 : Restrict an unconstrained granulated feature. If **S1** could not be applied we will try to resolve the conflict through restricting a previously unconstrained, granulated feature. We are therefore looking at the following features:

$$\forall i_{S2} : 1 \leq i_{S2} \leq n_{\text{granul}}$$
$$\wedge \forall c : 1 \leq c \leq C_{i_{S2}}^{\text{in}} \wedge s_{i_{S2}}^c = 1$$
$$\wedge \exists c : 1 \leq c \leq C_{i_{S2}}^{\text{in}} \wedge \mu_{i_{S2}}^{\text{in},c}(x) = 0 \wedge c \neq c_{\lambda,i_{S2}}$$

that is, all linguistic are contained in the constraint and we again make also sure that at least one linguistic value will remain in the modified constraint and we do not loose coverage of the anchor of this rule. Similar to above, if no such feature can be found, alternative **S3** is probed.

Otherwise we proceed analogous to scenario **S1**, that is, we compute the possible losses in volume and choose the one feature for shrinkage which results in the minimum loss in volume.

Example 4. The feature "temperature" of the rule from above can now not be restricted any further without eliminating the rule entirely.

$$\mathcal{R} : \text{if } temperature \text{ is } cold \text{ then } buy \ oil$$

Therefore the previously unconstrained feature "oil price" will be restricted and $\mathcal{R}$ might become:

$$\mathcal{R}' : \text{if } temperature \text{ is } cold \text{ and } oilprice \text{ is } low \text{ then } buy \ oil$$

S3 : Restrict an already constrained free feature. If none of the granulated features can be restricted to avoid the conflict, one of the remaining free features needs to be used. We will first try to tighten an already existing constraint on a free feature, that is we restrict ourselves to:

$$\forall i_{S3} : n_{\text{free}} + 1 \leq i_{S3} \leq n$$
$$\wedge \operatorname{cond}^{(r)}_{\text{free}, i_{S3}} = x_{i_{S3}} \text{ is } < a^{(r)}_{i_{S3}}, b^{(r)}_{i_{S3}}, c^{(r)}_{i_{S3}}, d^{(r)}_{i_{S3}} >$$

that is, all free features i for which a constraining trapezoidal membership function is already defined. If no such feature can be found, alternative **S4** is probed.

Otherwise – similar to cases **S1** and **S2** – we determine which of these features results in a minimum loss of rule-coverage. First for each feature i_{S3} we compute a modified trapezoid which eliminates the conflict:

$$< a'^{(r)}_{i_{S3}}, b'^{(r)}_{i_{S3}}, c'^{(r)}_{i_{S3}}, d'^{(r)}_{i_{S3}} >$$
$$:= \begin{cases} < x_{i_{S3}}, \max\{x_{i_{S3}}, b^{(r)}_{i_{S3}}\}, c^{(r)}_{i_{S3}}, d^{(r)}_{i_{S3}} > & : & x_{i_{S3}} < \lambda^{(r)}_{i_{S3}} \\ < a^{(r)}_{i_{S3}}, b^{(r)}_{i_{S3}}, \min\{x_{i_{S3}}, c^{(r)}_{i_{S3}}\}, x_{i_{S3}} > & : & x_{i_{S3}} > \lambda^{(r)}_{i_{S3}} \end{cases}$$

that is, depending on the position of the conflict with respect to the anchor $\lambda_{i_{S3}}$ the left or right side of the trapezoid is modified. This ensures that the anchor will always remain inside this rule but the conflict will be moved to the border of the support area.

Similar to **S1** we replace $\operatorname{cond}^{(r)}_{\text{free}, i_{S3}}$ using the new trapezoid and compute the loss in volume for the modified rule. One feature $i_{S3, \text{best}}$ will then be selected which minimizes this loss and the corresponding rule will be replaced in the set of rules $\mathbf{R}$.

Example 5.

$$\mathcal{R} : \text{if } temperature \text{ is } cold \text{ and } x_{42} \text{ is } < 2, 3, 7, 9 > \cdots$$

might become, through restriction of feature x_{42} to avoid a conflict at $x_{42} = 8$:

$$\mathcal{R}' : \text{if } temperature \text{ is } cold \text{ and } x_{42} \text{ is } < 2, 3, 7, 8 > \cdots$$

S4 : Restrict an unconstrained free feature. If no constrained free feature can be found, one of the unconstrained free feature needs to be constrained. That is we restrict ourselves to:

$$\forall i_{S4} : n_{\text{free}} + 1 \le i_{S4} \le n$$
$$\wedge \text{cond}^{(r)}_{\text{free}, i_{S4}} = \text{true}$$

that is, all free features i_{S4} for which a constraining trapezoidal membership function has not been defined. If no such feature can be found, a serious conflict has been encountered, that is, two example points have the same input vector x but conflicting targets μ. Usually this will result in a feedback to the user, pointing out this inconsistency in the data set. Otherwise – similar to case **S3** – we determine which of these features results in a minimum loss of rule-coverage. First for each feature i_{S4} we compute a new trapezoid which eliminates the conflict:

$$< a'^{(r)}_{i_{S4}}, b'^{(r)}_{i_{S4}}, c'^{(r)}_{i_{S4}}, d'^{(r)}_{i_{S4}} >$$
$$:= \begin{cases} < x_{i_{S4}}, \lambda^{(r)}_{i_{S4}}, \lambda^{(r)}_{i_{S4}}, +\infty > & : \quad x_{i_{S4}} < \lambda^{(r)}_{i_{S4}} \\ < -\infty, \lambda^{(r)}_{i_{S4}}, \lambda^{(r)}_{i_{S4}}, x_{i_{S4}} >, & : \quad x_{i_{S4}} > \lambda^{(r)}_{i_{S4}} \end{cases}$$

that is, depending on the position of the conflict with respect to the anchor $\lambda_{i_{S4}}$ the left or right side of the trapezoid is restricted. This ensures that the anchor will always remain inside this rule but the conflict x_i will be moved to the border of the support area. The core itself is set equal to the anchor λ_{S4} and will later be enlarged if this rule covers other patterns.

Similar to **S3** we then replace $\text{cond}^{(r)}_{\text{free}, i_{S4}}$ and compute the loss in volume for the modified rule. Feature $i_{S4,\text{best}}$ will then be selected to minimize this loss and the respective rule will be replaced in the set of rule **R**.
Example 6.

$$\mathcal{R} : \text{if } temperature \text{ is } cold \cdots$$

might become (through restriction of the previously unconstrained feature x_{42}):

$$\mathcal{R}' : \text{if } temperature \text{ is } cold \text{ and } x_{42} \text{ is } < -\infty, 3, 3, 7 > \cdots$$

if $\lambda_{42} = 3$, and the conflict occurred at $x_{42} = 7$.

3   Two Stage Fuzzy Models

Most existing algorithms to construct fuzzy rule sets from data have tremendous problems with noisy data or data containing outliers. Usually an excessive number of rules is being introduced simply to model noise and/or outliers. This also applies to the algorithm described in the previous section. This is due to the fact that these algorithms aim to generate conflict free rules, that is, examples encountered during training will result in a degree of membership > 0 only for those rules of the correct class (or granule). Unfortunately in case of spares outliers such an approach will, especially in high-dimensional feature spaces, result in an enormous amount of rules to avoid these conflicts. Figure 2 demonstrates this effect.

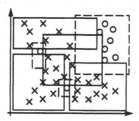

Fig. 2. An example how two outliers (o) produce a high number of rules for the competing class ×.

Using the already existing model we can, however, in many cases easily determine parts that have low relevance, based on their weight or another parameter which denotes individual relevance. To measure a rule's relevance often the weight parameter $w^{(r)}$ is used which represents the number of training patterns covered by rule r. From this a measure for the importance or relevance of each rule can be derived, by simply using the percentage of patterns covered by this rule:

$$\Phi(\mathcal{R}^{(r)}) = \frac{w^{(r)}}{|\mathbf{T}|}.$$

Other measures which are also used determine the loss of information if rule r is omitted:

$$\Phi(\mathcal{R}^{(r)}) = I(\mathbf{R}) - I(\mathbf{R}\backslash\{\mathcal{R}^{(r)}\})$$

where $I(\cdot)$ indicates a function measuring the information content of a rule set. Most commonly used are (a more extensive overview can be found in [3] and also [1]):

- the Gini-index:

$$I_{\text{Gini}}(\mathbf{R}) = 1 - \sum_{i=1}^{C^{\text{out}}} V_i(\mathbf{R})^2,$$

and the fuzzy entropy function:

$$I_{\mathbf{Entropy}}(\mathbf{R}) = -\sum_{i=1}^{C^{\mathrm{out}}} (V_i(\mathbf{R}) \log_2 V_i(\mathbf{R}))$$

where $V_i(\mathbf{R})$ indicates the volume of all rules $\mathcal{R}^{(r)}$ in $\mathbf{R}$ which are assigned to granule $i = g^{(r)}$:

$$V_i(\mathbf{R}) = \int_{\mathbf{x}} \max_{\mathcal{R}^{(r)} \in \mathbf{R} \wedge g^{(r)} = i} \{\mu^i(\mathbf{x})\} d\mathbf{x}$$

(In [9] it is shown how this volume can be computed efficiently based on a system of fuzzy rules.)

The choice of relevance-measure is made depending on the nature of the underlying rule generation algorithm, as well as the focus of analysis, i.e. the interpretation of important vs. unimportant or useless data points. Using such a measure of (notably subjective) relevance, we can now extract rules with low relevance from this model, assuming that they describe points in the data which are outliers or spares points:

$$\mathbf{R}_{\mathrm{outlier}} = \mathbf{R} \backslash \{\mathcal{R} \in \mathbf{R} \mid \Phi(\mathcal{R}) < \theta_{\mathrm{outlier}}\}$$

Using this "outlier"-model as filter for a second training phase will then generate a new fuzzy model which has less rules with higher significance. In fact, the original training data $\mathbf{T}$ is filtered and only data points which are not covered by the outlier model will be used to construct the new model:

$$\mathbf{T}_{\mathrm{clean}} = \{(\mathbf{x}, \mu^{\mathrm{target}}) \in \mathbf{T} \mid \forall \mathcal{R} \in \mathbf{R}_{\mathrm{outlier}} : \mu_{\mathcal{R}}(\mathbf{x}) \leq \theta_{\mathrm{filter}}\}$$

Figure 3 shows the flow of this procedure. Note, how the initial model is being used to extract the outlier-model. This model is then in turn used as a filter for the existing training data to generate the final model. In Figure 4 (left) the outlier model is shown, covering the two single points of class o. Now the

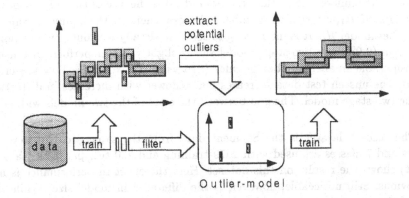

Fig. 3. The role of the two models during training.

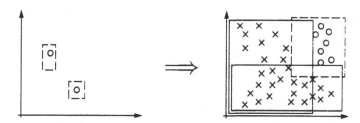

Fig. 4. The same example, but here the additional outlier-model (left) reduces the number of generated rules in the second stage model (right).

second phase can ignore these points and generate two large rules describing the remaining data points of class $\times$ (right). In the following section we will show how this effects the size of the rule sets on two real-world datasets.

4   Experiments

Experiments on two datasets from the Statlog-archive [8] were performed to demonstrate the effect of the proposed methodology. The relevance function $\Phi(R^{(j)}) = w^{(j)}$ with a threshold of $\theta = 5$ was used, that is rules which cover less then five patterns were considered irrelevant.

The first dataset contains images from Satellites (Satimage-dataset), patterns with 36 attributes have to be separated into 6 different classes. All together 4435 training and 2000 test patterns were used. Table 1 (left) shows the results. Here "Standard" stands for the normal algorithm which generates fuzzy rules in one run. $\mathcal{H}_1$ indicates the general model generated through the algorithm explained above. The number of rules for both models is shown in the last column. The number before the brackets indicates the size of the rule-set of the general model $\mathcal{H}_1$, whereas the number in brackets denotes the number of rules of the outlier model $\mathcal{H}_0$. It is interesting to see how already without any additional distortion (0.0%) the two-stage model shows slightly better performance using a considerable smaller number of rules (270. vs. 393). Note also how the error rate on the unseen test data increases much slower with increases in distortion for the two-stage model. The gap between the sizes of the two models widens as well.

The second dataset is the Segment data from the same archive. Here 19 inputs and 7 classes are used with 2079 training and 220 test patterns. Table 1 (right) shows the results on this dataset. Here the effect in performance is not as obvious. Still noticeable, however, is the difference in model size. While the size of the separate outlier-model increases with increasing distortion, the size of the model representing the more general behavior grows much slower.

Table 1. Results on the Satimage (left) and Segment (right) dataset

level of distort.	used model	error on test data	#rules		level of distort.	used model	error on test data	#rules
0.0%	Stand.	15.9%	393		0.0%	Stand.	3.5%	96
	$\mathcal{H}_1(\mathcal{H}_0)$	13.5%	270 (60)			$\mathcal{H}_1(\mathcal{H}_0)$	3.0%	80 (12)
1.0%	Stand.	17.1%	394		1.0%	Stand.	5.2%	108
	$\mathcal{H}_1(\mathcal{H}_0)$	13.5%	313 (81)			$\mathcal{H}_1(\mathcal{H}_0)$	4.3%	86 (22)
2.0%	Stand.	18.1%	404		2.0%	Stand.	6.9%	113
	$\mathcal{H}_1(\mathcal{H}_0)$	12.9%	295 (109)			$\mathcal{H}_1(\mathcal{H}_0)$	5.6%	83 (30)
5.0%	Stand.	18.1%	479		5.0%	Stand.	6.1%	144
	$\mathcal{H}_1(\mathcal{H}_0)$	12.4%	334 (145)			$\mathcal{H}_1(\mathcal{H}_0)$	3.8%	107 (37)
10.0%	Stand.	22.3%	578		10.0%	Stand.	6.5%	151
	$\mathcal{H}_1(\mathcal{H}_0)$	15.2%	379 (199)			$\mathcal{H}_1(\mathcal{H}_0)$	6.5%	106 (45)

5 Conclusions

In this paper we discussed a strategy to model potential outliers through an additional outlier-model. This results in two models, one representing a more general concept which offers better understandability, and the other concentrating on potential outliers or regions with low evidence in the observed data. In the future an entire hierarchy of fuzzy models seems to be a promising approach to model large amounts of data and enable the user to investigate the underlying behavior at various levels of granularity, following Lotfi Zadeh's concept of "information granulation".

References

1. C. Apte, S. Hong, J. Hosking, J. Lepre, E. Pednault, and B. K. Rosen. Decomposition of heterogeneous classification problems. *Intelligent Data Analysis*, 2(2), 1998. (http://www.elsevier.nl/locate/ida).
2. M. R. Berthold and K.-P. Huber. Constructing fuzzy graphs from examples. *Intelligent Data Analysis*, 3(1), 1999. (http://www.elsevier.nl/locate/ida).
3. V. Cherkassky and F. Mulier. *Learning from Data*. John Wiley and Sons Inc., 1998.
4. S. K. Halamuge and M. Glesner. FuNe Deluxe: A group of fuzzy-neural methods for complex data analysis problems. In *Proceedings of the EUFIT'95*, Aug. 1995.
5. D. Hand, J. Kok, and M. Berthold, editors. *Advances in Intelligent Data Analysis*. LNCS. Springer Verlag, 1999.
6. C. M. Higgins and R. M. Goodman. Learning fuzzy rule-based neural networks for control. In *Advances in Neural Information Processing Systems*, 5, pages 350–357, California, 1993. Morgan Kaufmann.
7. K.-P. Huber and M. R. Berthold. Building precise classifiers with automatic rule extraction. In *IEEE International Conference on Neural Networks*, 3, pages 1263–1268, 1995.
8. D. Michie, D. J. Spiegelhalter, and C. C. Taylor, editors. *Machine Learning, Neural and Statistical Classification*. Ellis Horwood Limited, 1994.

9. R. Silipo and M. R. Berthold. Discriminative power of input features. In *[5]*, 1999.
10. P. K. Simpson. Fuzzy min-max neural networks – part 1: Classification. *IEEE Transactions on Neural Networks*, 3(5):776–786, Sept. 1992.
11. P. K. Simpson. Fuzzy min-max neural networks – part 2: Clustering. *IEEE Transactions on Fuzzy Systems*, 1(1):32–45, Jan. 1993.
12. L.-X. Wang and J. M. Mendel. Generating rules by learning from examples. In *International Symposium on Intelligent Control*, pages 263–268. IEEE, 1991.
13. L.-X. Wang and J. M. Mendel. Generating fuzzy rules by learning from examples. *IEEE Transactions on Systems, Man, and Cybernetics*, 22(6):1313–1427, 1992.
14. L. A. Zadeh. Soft computing and fuzzy logic. *IEEE Software*, pages 48–56, Nov. 1994.
15. L. A. Zadeh. Fuzzy logic and the calculi of fuzzy rules and fuzzy graphs: A precis. *Multi. Val. Logic*, 1:1–38, 1996.

AN ALGORITHM FOR ADAPTIVE CLUSTERING AND VISUALISATION OF HIGHDIMENSIONAL DATA SETS

F. Schwenker

University of Ulm, Ulm, Germany

H.A. Kestler

University of Ulm, Ulm, Germany
and
University Hospital Ulm, Ulm, Germany

G. Palm

University of Ulm, Ulm, Germany

Abstract

We describe an algorithm for exploratory data analysis which combines adaptive c-means clustering and multi-dimensional scaling (ACMDS). ACMDS is an algorithm for the online visualization of clustering processes and may be considered as an alternative approach to Kohonen's self organizing feature map (SOM). Whereas SOM is a heuristic neural network algorithm, ACMDS is derived from multivariate statistical algorithms. The implications of ACMDS are illustrated through five different data sets.

1 Introduction

In many practical applications one has to explore the underlying structure of a large set objects. Typically, each object is represented by a feature vector $x \in \mathcal{X}$, where $\mathcal{X}$ is the feature space endowed with a distance measure $d_\mathcal{X}$. This data analysis problem can be tackled utilizing *clustering methods* (see [Jain and Dubes, 1988] for an overview). A widely used clustering algorithm is *c-means clustering* [MacQueen, 1967, Lloyd, 1982, Linde et al., 1980] where the aim is to reduce a set of M data points $X = \{x_1, \ldots, x_M\} \subset \mathcal{X}$ into a few, but representative, cluster centers $\{c_1, \ldots, c_k\} \subset \mathcal{X}$.

A neural network algorithm for clustering is Kohonen's *selforganizing feature map (SOM)* [Kohonen, 1995]. SOM is similar to the classical sequential c-means algorithm (see section 2) with the difference that in SOM the cluster centers are mapped into a *display space* $\mathcal{Z}$ with a distance measure $d_\mathcal{Z}$. Each cluster center is mapped to a fixed location of the display space. The idea of this display space $\mathcal{Z}$ is that cluster centers corresponding to nearby points in $\mathcal{Z}$ have nearby locations in the feature space $\mathcal{X}$. Typically, $\mathcal{Z}$ is a 2-dimensional (or 3-dimensional) grid. Therefore it is often emphasized that Kohonen's SOM is able to combine clustering and visualization aspects. In this context it has to be mentioned that SOM is a heuristic algorithm — not derived from an objective function which incorporates both a clustering criterion and some kind of topological or neighborhood preserving measure [Kohonen, 1995, Ritter and Schulten, 1988].

Another approach for getting an overview over a high-dimensional data set $X \subset \mathcal{X}$ are *visualization methods*. In multivariate statistics several linear and nonlinear techniques have been developed. A widely used nonlinear visualization method is *multidimensional scaling (MDS)* [Scott, 1992, Sammon, 1969]. MDS is a class of distance preserving mappings from the data set X into a low-dimensional *projection space* $\mathcal{Y}$ which is endowed with some distance measure $d_\mathcal{Y}$. Each feature vector $x_\mu \in X$ is mapped to a point $y_\mu := p(x_\mu) \in \mathcal{Y}$ in such a way that the distance matrix $D_\mathcal{X} := (d_\mathcal{X}(x_i, x_j))_{1 \leq i,j \leq M}$ in feature space $\mathcal{X}$ is approximated by the distance matrix $D_\mathcal{Y} := (d_\mathcal{Y}(y_i, y_j))_{1 \leq i,j \leq M}$ in projection space $\mathcal{Y}$.

In this paper an algorithm, which we call ACMDS, is described. It combines c-means clustering and MDS. This procedure is able to be utilized for the online visualization of clustering processes and should be considered as an alternative approach to Kohonen's self organizing feature (SOM).

Throughout this paper we restrict our considerations to the feature space $\mathcal{X} = \mathbb{R}^n$ and the projection space $\mathcal{Y} = \mathbb{R}^r$. The Euclidean distance d is used as distance measure in both spaces.

The paper is organized as follows. The three basic algorithms *c-means*, *SOM*, and *MDS* are discussed in section 2. In section 3 we describe the *ACMDS* algorithm; which is then applied to five different data sets two artifical and three real world data sets. We discuss these numerical experiments in section 4.

2   Components of ACMDS

The *c-means clustering* algorithm moves a fixed set of k cluster centers into the centers of gravity of the accumulations of data points [MacQueen, 1967, Lloyd, 1982]. For the interpretation of the clustering result it is important to choose the right number of cluster centers k. If the choosen number of clusters is different from the actual number of clusters hidden within the data, the result of the clustering process has to be reconsidered.

During the c-means clustering process the current data point $x \in X$ is classified to the closest center c_{j^*}, i.e. the data point x belongs to cluster C_{j^*} if $d(x, c_{j^*}) = \min_i d(x, c_i)$. The quantization error $H(c_1, \ldots, c_k)$ defined by

$$H(c_1, \ldots, c_k) = \sum_j \sum_{x \in C_j} d^2(x, c_j)$$

is minimial, if each cluster center c_j is equal to the corresponding center of gravity of cluster C_j, e.g. if for all $j = 1, \ldots, k$ hold

$$c_j = \frac{1}{|C_j|} \sum_{x \in C_j} x$$

where $|C_j|$ denotes the size of cluster C_j. This type of algorithm is called *batch c-means* [MacQueen, 1967].

We concentrate on the *sequential c-means clustering* method realized by the updating rule

$$\Delta c_{j^*} = \frac{1}{|C_{j^*}| + 1}(x - c_{j^*}) \tag{1}$$

where c_{j*} is the closest cluster center to the data point $x \in X$ [Lloyd, 1982].

Sequential c-means clustering is closely related to learning in artificial neural networks in particular, to the paradigm of *competitive neural networks* [Moody and Darken, 1989]. *Kohonen's selforganizing feature map (SOM)* is a competitive learning scheme [Kohonen, 1995]. During SOM learning cluster centers c_j that are close in the display space $\mathcal{Z}$ will be adapted to the same input $x \in X$:

$$\Delta c_j = \eta_t h(r_j, r_{j*})(x - c_j) \tag{2}$$

where $h : \mathcal{Z} \times \mathcal{Z} \to \mathbb{R}_+$ is a neighborhood function with $h(r_j, r_{j*}) \to 0$ for increasing distance of r_j and r_{j*}. For the special case of $h(r_j, r_{j*}) = 1$ for $j = j^*$ and $h(r_j, r_{j*}) = 0$ otherwise, Kohonen's SOM updating rule is identical to sequential c-means clustering.

Given a set of data points $X \subset \mathcal{X}$ and a transformation $p : \mathcal{X} \to \mathcal{Y}$ a natural evaluation criterion for the distance preservation of p is some kind of difference between the matrices $D_{\mathcal{X}} := (d_{\mathcal{X}}(x_i, x_j))_{1 \leq i, j \leq M}$ in $\mathcal{X}$ and $D_{\mathcal{Y}} := (d_{\mathcal{Y}}(y_i, y_j))_{1 \leq i, j \leq M}$ in $\mathcal{Y}$.

MDS is a multivariate statistics procedure that start with a distance matrix $D_{\mathbb{R}^n}$ of M data points $X = \{x_1, \ldots, x_M\} \subset \mathbb{R}^n$ and generates a set of corresponding representation points $Y = \{p(x_1), \ldots, p(x_M)\} \subset \mathbb{R}^r$ [Jain and Dubes, 1988, Sammon, 1969]. This projection $p : X \to \mathbb{R}^r$ is calculated in such a way that the distance matrices $D_{\mathbb{R}^n}$ and $D_{\mathbb{R}^r}$ are similar. The difference is measured through *stress functions* defined by

$$S(x_1, \ldots, x_M) = \alpha \sum_{j,i=1}^{M} \left(\Phi[d^2(x_i, x_j)] - \Phi[d^2(y_i, y_j)] \right)^2 \tag{3}$$

where $\alpha > 0$ and $\Phi : \mathbb{R} \to \mathbb{R}$ is a strictly increasing differentiable function, i.e. $\Phi(x) = x$, $\Phi(x) = \sqrt{x}$ or $\Phi(x) = \log(x + 1)$. The consecutive points y_j may be calculated by a gradient descent algorithm

$$\Delta y_j = \eta_t \cdot \alpha \sum_{i \neq j}^{M} \delta_{ji}(y_i - y_j), \tag{4}$$

where δ_{ji} is the weighted difference between $d(x_i, x_j)$ and $d(y_i, y_j)$ given by

$$\delta_{ji} = \Phi'[d^2(y_i, y_j)] \left(\Phi[d^2(x_i, x_j)] - \Phi[d^2(y_i, y_j)] \right)$$

and $\eta_t > 0$ with $\eta_t \to 0$ as $t \to \infty$.

3 The ACMDS–Algorithm

In this approach we combine the sequential c-means clustering procedure, to detect the cluster structure in feature space, with a MDS algorithm, to get a low-dimensional representation of the cluster centers. To achieve this, the cluster centers $c_j \in \mathbb{R}^n$ move according to the sequential c-means iteration rule

(1) and simultanously a set of low-dimensional representation centers $p_j :=$ $p(c_j)$ move in $\mathbb{R}^r$. These representation centers move in such a way that the distances $d(p_i, p_j)$ are close to the distances $d(c_i, c_j)$ of the cluster centers. This is realized by a gradient descent algorithm,

$$\Delta p_j = \eta_t \alpha \sum_{i \neq j}^{k} \delta_{ji}(p_i - p_j) \tag{5}$$

where δ_{ji} is

$$\delta_{ji} = \Phi'[d^2(p_i, p_j)]\Big(\Phi[d^2(c_i, c_j)] - \Phi[d^2(p_i, p_j)]\Big)$$

which minimizes the stress function

$$S(p_1, \ldots, p_k) = \alpha \sum_{i,j=1}^{k} \Big(\Phi[d^2(c_i, c_j)] - \Phi[d^2(p_i, p_j)]\Big)^2 \tag{6}$$

ACMDS algorithm

```
estimate thresholds θ_new and θ_merge
set k = 0 (no prototypes)
        choose a data point x ∈ X
        calculate d_j = d(x, c_j), j = 0, ..., k
        detect the winner j* = argmin_j d_j
        if (d_j* > θ_new) or k = 0
            c_k := x and p_k according to (5)
            k := k + 1
        else
            adapt c_j* by (1) and p_j* by (5)
            calculate D_l = d(c_l, c_j*), l = 0, ..., k
            detect l* := argmin_{l≠j*} D_l
            if (D_l* ≤ θ_merge)
                merge(c_l*, c_j*), k := k - 1
        goto: choose data point
```

In addition we incorporate into the clustering algorithm a scheme for adjusting the number k of cluster centers in order to address the problem of finding a good choice for number of cluster centers and initial locations of cluster centers. We use a variant of c-means clustering in the ACMDS procedure, which allows the creation of new cluster centers and the merging of close clusters and call it *adaptive c-means clustering*. For this clustering algorithm parameters θ_{merge} and θ_{new} have to be derived from the data subset X in advance.

When a new cluster center is inserted, the location of the corresponding representation center p_j has to be determined, by setting the inital position to be a linear combination of its two nearest neighbors and then adapting p_j by the iteration rule (5).

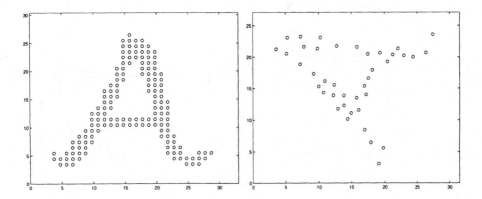

Figure 1: The A-shaped data set X with 153 data points (left) and a set of 40 projection centers (right) are shown (see text).

4   Numerical Results and Discussion

To illustrate the behavior of the ACMDS algorithm we present results for five different data sets. They vary in size and dimension of the feature space.

4.1   Artifical Data Set

In Figure 1 a twodimensional data set X, containing 153 samples ist shown. This A-shaped data set was made by hand using a standard icon editor. This set X was transformed by a rotation $\phi_\alpha : \mathbb{R}^2 \to \mathbb{R}^2$ with a randomly choosen angle $\alpha \in (0, 2\pi)$ onto X_α. Then, X_α was embbed into $\mathbb{R}^{128}$ by the mapping $\phi_{ij} : \mathbb{R}^2 \to \mathbb{R}^{128}$. This is defined by $x = (x_1, x_2) \mapsto z = (z_1, \ldots, z_{128})$ where $z_i := x_1$, $z_j := x_2$ and $z_k := 0$ for all k different form i and j. Obviously, the Euclidean distances between the data points $\{x_\mu\}$ and $\{z_\mu\}$ were not changed by these two transformations. After that, each component i of the data points $\{z_\mu\}$ was corrupted by a small amount of Gaussian random noise: $z_i^\mu := z_i^\mu + 0.1\mathcal{N}(0, 1)$. This set $\{z^\mu\}$ was analyzed by the ACMDS algorithm. In Figure 1 (right) the set of 40 projection centers $\{p_j\}$, also A-shaped as the set X, is given.

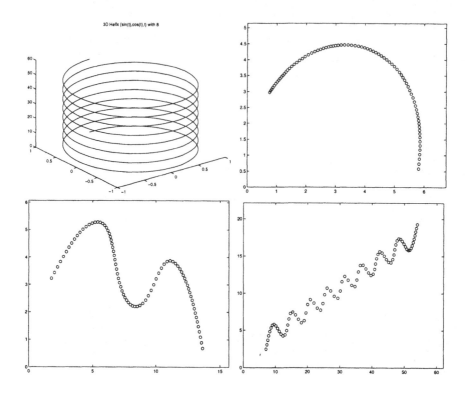

Figure 2: The 3D-helix $H_8 := \{(sin(t), cos(t), t) \mid 0 \le t \le 16\pi\}$ and sets of 2D-projection centers for three finite subsets of H_1, H_2 and H_8 (see text).

4.2 3D-Helix

Next we give some results of subsets of the 3-dimensional helix

$$H := \{(sin(t), cos(t), t) \mid t \ge 0\} \subset \mathbb{R}^3.$$

In Figure 2 the data set H_8, containing the first 8 loops of H is shown. From the data set H_k finite subsets X_k were equidistantly sampled with approximately 500 sample points per loop. Four data sets X_k, with $k = 1, 2, 4, 8$, were analyzed by the ACMDS algorithm. For the data set X_1, X_2 and X_8 sets of representation centers $p_j \in \mathbb{R}^2$ are given, see Figure 2. It can be observed that the data set X_k is projected onto a sine-like curve with approximately k periods.

Figure 3 shows the clustered and projected data set X_4. From the 2000 samples of X_4 80 cluster centers are projected. Snapshots after $500, 1000, 1500$ and 3000 adaptation steps are given.

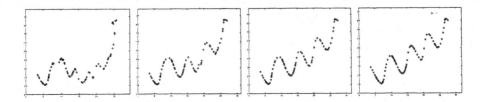

Figure 3: The projection centers p_j of the data set X_4 (see text) calculated by ACMDS after $500, 1000, 1500$ and 3000 online adaptation steps.

4.3 Data Set of Handwritten digits

The data set used in this application consisted of 10.000 handwritten digits with 1000 samples per class. The digits were normalized in height and width. Each digit is represented by a 16×16 matrix (g_{ij}) where $g_{ij} \in \{0, \ldots, 255\}$ is a value from a 8 bit gray scale (for details concerning the data set see [Kreßel, 1991]). Figure 4 shows 10 samples from this data base and a subset of 10 cluster centers calculated by the adaptive clustering algorithm as gray scale images. The whole set of projection centers of this data set is shown as a 2D-map in Figure 5. Here, each projection center p_j is labeled by a class label k and frequency $freq \in [0, 1]$. Denoting the Voroni region of cluster center c_j with $V_j := \{x \in \mathbb{R}^n : d(x, c_j) = \min d(x, c_k)\}$, and $X^l := \{x \in X : x$ is from class $l\}$ for $l = 0, \ldots, 9$ then the class label k and the frequency $freq$ are defined by:

$$|V_j \cap X_k| = \max\{V_j \cap X_l : l = 0, \ldots, 9\} \quad \text{and} \quad freq = \frac{|X_k \cap V_j|}{|X \cap V_j|}.$$

Figure 4: In the first row handwritten digits from the data set of 10000 patterns are shown. A subset of 10 cluster centers calculated by the adaptive clustering algorithm is shown as gray scale images in the second row.

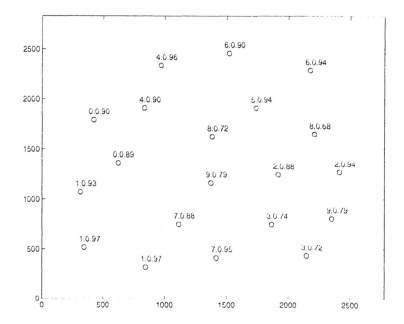

Figure 5: The whole set of projection centers p_j of the data set of handwritten digits is shown as a 2D-map, where each p_j is labeled by $k : freq$, with a class label k and a frequency $freq$. Label k is defined as the class which labels the majority of data points $x \in X \cap V_j$. Frequency $freq$ stands for the fraction of data points from this majority class k.

4.4   Data Set of Highresolution ECG's

The next data set stems from the problem domain of classifying ECG signals with the purpose of predicting sudden cardiac death.

Background

The incidence of sudden cardiac death (SCD) in the area of the Federal Republic of Germany is about 100.000 to 120.000 cases per year. Studies showed that the basis for a fast heartbeat which evolved into a heart attack is a localized damaged heart muscle with abnormal electrical conduction characteristics. These conduction defects, resulting in an abnormal contraction of the heart muscle may be monitored by voltage differences of electrodes fixed to the chest. This is the electrocardiogram (ECG) which reflects the conduction characteristics of the heart.

Damaged regions within the heart have been explored with microelectrodes and have been found to contain irregular conducting cells. This causes slow or irregular propagation of activation. In the electrocardiogram this is represented through waves that can extend beyond one heartbeat. The ECG is a save method to obtain information about the heart, i.e. there is no risk for the patient of recording an ECG. Consequently it is very desireable to use the ECG as a screening method to categorize subjects according to their risk of SCD.

High-resolution electrocardiography is used for the detection of fractionated micropotentials, which serve as a noninvasive marker for an arrhythmogenic substrate and for an increased risk for malignant ventricular tachyarrhythmias. Ventricular late potential analysis (VLP) is herein the generally accepted noninvasive method to identify patients with an increased risk for reentrant ventricular tachycardias and for risk stratification after myocardial infarction [Simson, 1981, Gomes et al., 1989, Höher and Hombach, 1991a, Höher and Hombach, 1991b]. Techniques commonly applied in this purely time–domain based analysis are signal-averaging, high-pass filtering and late potential analysis of the terminal part of the QRS complex. The assessment of VLP's depends on three empirically defined limits of the total duration of the QRS and the duration and amplitude of the terminal low–amplitude portion of the QRS [Breithardt and Borggrefe, 1986, Breithardt et al., 1991].

Subject Data and Recordings

High resolution beat-to-beat recordings were obtained from 95 subjects separated into two groups: Group A consisted of 51 healthy volunteers (age 24±4.2 years, range 16-34; 31 men and 20 women) without any medication. In order to qualify as healthy, the following risk factors and illnesses had to be excluded: angina pectoris, dyspnoe, dizzyness, syncope arrhythmias, rhymatic fever, diphteria, myocarditis, influenza within the last four weeks and the cardiac risk factors hyperlipidaemia, high blood pressure, diabetes, smoking (more than 5 cigarettes a day) and the contraceptive pill. In family history (relatives

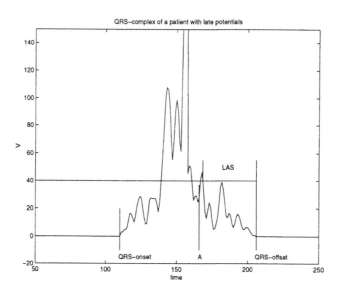

Figure 6: QRS-complex of a patient with "Late Potentials". The time A is defined as $A := QRS_{offset} - 40ms$.

once or twice removed) there had to be excluded myocardial infarction and sudden cardiac death under the age of 60. On examination there had to be: normal blood pressure, sinus rhythm, no conduction defect in the 12 channel ECG, normal dimensions of all four chambers with normal left ventricular function and normal regional wall motion of the left and right ventricular detected by echocardiography. Subjects older than 30 years had to have a normal exercise ECG. For these 51 subjects 68 were examined. Exlusion critera were right bundle branch block (3), influenza in the last 4 weeks (2), pericardial effusion (1), abnormal blood pressure (4), myocardial infarction of relatives once or twice removed (2), first degree atrioventricular block (1) and smoking more than 5 cigarettes a day (4). In the 51 included subjects there were 7 smoking less than 5 cigarettes a day, there was one borderline left ventricular dilation which was an 'athletes heart' and one rr'-phenomenon in ECG.

Group B consisted of 44 patients. Inclusion criteria were an inducible clinical ventricular tachycardia (>30 sec) at electrophysiologic study (EPS) with a history of myocardial infarction and coronary artery disease in angiogram. All patients had to be without antiarrhythmic medication. In ECG there had to be sinus rhythm without any conduction defect. Exclusion criteria were a congestive or hypertrophic cardiomyopathy, preexitation syndrome, long QT-syndrome or arrhythmogenic right ventricular dysplasia and electrolyte derangements. For these 44 patients 461 patients with an EPS were screened. All patients had a hypokinesia or an akinesia in the infarct related myocardium, 30 patients additionally showed an abnormal wall motion in non-infarct related areas documented either by laevocardiography or echocardiography. Median duration between last myocardial infarction and EPS was 1242 days (range 7-34388).

Signal-averaged high resolution ECGs (see Figure 6) were recorded from three orthogonal bipolar leads (sampling rate 2000 Hz and A/D-resolution was 16 bits filtered with 40–250 Hz bidirectional 4-pole Butterworth filter).

Features from the vector magnitude $V = \sqrt{X^2 + Y^2 + Z^2}$ are:

- Total duration of the filtered QRS:

$$QRSD := QRS_{offset} - QRS_{onset}$$

- RMS of the terminal 40 ms of the QRS:

$$tRMS := \sqrt{\frac{1}{QRS_{offset} - A} \sum_{i=A}^{QRS_{offset}} V_i^2}$$

- Terminal low-amplitude signal of the QRS below 40 μV:

$$LAS := QRS_{offset} - argmax\{i \mid V_i \geq 40\mu V\}$$

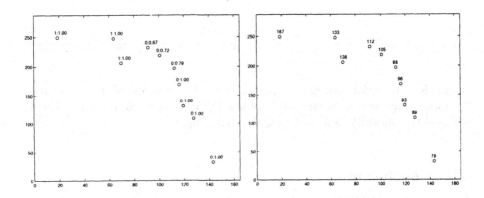

Figure 7: Projection centers calculated from three features $(QRSD, tRMS, LAS)$ of the QRS-complex. Annotation: Class labels of the majority (left figure); QRS_d (right figure).

First, results of the clustering and visualization procedure are shown for the three features $(QRSD, tRMS, LAS)$ in Figure 7. In the left part of Figure 7, each projection center is labeled with the class label (coronary heart disease =1; healthy =0) of the majority and the fraction of data points within this cluster (cf. section 4.3). In the right 2D-map of Figure 7 the same representation centers p_j are shown, but now labeled with the mean of the QRS-duration.

It can be observed that the set of representation centers is grouped around a 1-dimensional L-shaped feature space and that the chain of representation centers is ordered by the $QRSD$-labels.

These results on signal averaged ECG's (Figure 7) show that conventional analysis is dominated by the feature QRS-duration, i.e the averaged length of the duration of repolarization of the heart muscle. This is very apparent in Figure 7, surprisingly an ordering of the classes by the QRS-duration is also visible in Figure 8. This is quite astonishing, as the QRS-duration is not a feature in this data. Here, the terminal 50 msec ($= 101$ sample points) of the vector magnitude signal V (see Figure 6) of the QRS-complex were used as input data. These findings support the assumption of an identical mechanism influencing the terminal part of the QRS as well as the QRS-duration in terms of inducible ventricular arrhythmias.

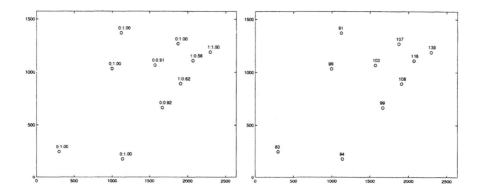

Figure 8: Projection centers calculated from the terminal 50 msec ($= 101$ samples) of the vector magnitude V of the QRS-complex. Annotation: Class labels of the majority (left figure); QRS_d (right figure).

4.5   Conclusion

The presented ACMDS-algorithm is an adaptive data analysis procedure for clustering and visualization of large and high dimensional data sets. The adaptivity of this procedure makes it useful for many applications, where the clustering itself is part of a larger program operating in an enviroment under human supervision. Furthermore, it may help in forming hypotheses about the data, which in turn may be substantiated by other statistical methods.

Acknowledgment

This work was partly supported by the collaborative research center (SFB 527) at the University of Ulm.

References

[Breithardt and Borggrefe, 1986] Breithardt, G. and Borggrefe, M. (1986). Pathophysiological mechanisms and clinical significance of ventricular late potentials. *Eur Heart J*, 7:364–385.

[Breithardt et al., 1991] Breithardt, G., Cain, M., El-Sherif, N., Flowers, N., Hombach, V., Janse, M., Simson, M., and Steinbeck, G. (1991). Standards for analysis of ventricular late potentials using high resolution or signal-averaged electrocardiography. *Eur Heart J*, 12:473–80.

[Gomes et al., 1989] Gomes, J., Winters, S., Martinson, M., Machac, J., Stewart, D., and Targonski, A. (1989). The prognostic significance of quantitative signal-averaged variables relative to clinical variables, site of myocardial infarction, ejection fraction and ventricular premature beats. *JACC*, 13:377–384.

[Höher and Hombach, 1991a] Höher, M. and Hombach, V. (1991a). Ventrikuläre Spätpotentiale – Teil I Grundlagen. *Herz & Rhythmus*, 3(3):1–7.

[Höher and Hombach, 1991b] Höher, M. and Hombach, V. (1991b). Ventrikuläre Spätpotentiale – Teil II Klinische Aspekte. *Herz & Rhythmus*, 3(4):8–14.

[Jain and Dubes, 1988] Jain, A. and Dubes, R. (1988). *Algorithms for Clustering Data*. Prentice Hall, Englewood Cliffs, New Jersey.

[Kohonen, 1995] Kohonen, T. (1995). *Self-Organizing Maps*. Springer.

[Kreßel, 1991] Kreßel, U. (1991). The Impact of the Learning-Set Size in Handwritten-Digit Recognition. In Kohonen, T., editor, *Artificial Neural Networks*. ICANN-91, North-Holland.

[Linde et al., 1980] Linde, Y., Buzo, A., and Gray, R. (1980). An algorithm for vector quantizer design. *IEEE Transactions on Communications*, 28(1):84–95.

[Lloyd, 1982] Lloyd, S. (1982). Least squares quantization in PCM. *IEEE Transactions on Information Theory*, 28(2):129–137.

[MacQueen, 1967] MacQueen, J. (1967). Some methods for classification and analysis of multivariate observations. In L.M.LeCam and J.Neyman, editors, *Proceedings of the Fifth Berkeley Symposium on Mathematical Statistics and Probability*, volume I, pages 281–297. Berkeley University of California Press.

[Moody and Darken, 1989] Moody, J. and Darken, C. (1989). Fast learning in networks of locally-tuned processing units. *Neural Computation*, 1(2):281–294.

[Ritter and Schulten, 1988] Ritter, H. and Schulten, K. (1988). Convergence properties of Kohonen's topology converving maps:fluctuations,stability, and dimension selection. *Biological Cybernetics*, 60:59–71.

[Sammon, 1969] Sammon, J. (1969). A nonlinear mapping for data structure analysis. *IEEE Transactions on Computers*, C-18:401–409.

[Scott, 1992] Scott, D. (1992). *Multivariate Density Estimation*. John Wiley & Sons, New York.

[Simson, 1981] Simson, M. (1981). Use of Signals in the Terminal QRS Complex to Identify Patients with Ventricular Tachycardia after Myocardial Infarction. *Circulation*, 64(2):235–242.

LEARNING IN COMPUTER SOCCER

H.-D. Burkhard

Humboldt University Berlin, Berlin, Germany

Abstract: Computer Soccer is a testbed for intelligent autonomous machines and programs under real life conditions. Besides others, it provides challenging problems in the design of intelligent agents and in the field of machine learning.

1 Introduction

After the success of Deep Blue in chess, football playing robots are a new challenge for the development of intelligent machines [1]. Research and competitions are coordinated and organized by the RoboCup Federation:

> The Robot World Cup Initiative (RoboCup) is an attempt to foster AI and intelligent robotics research by providing a standard problem where a wide range of technologies can be integrated and examined. For this purpose, RoboCup chose to use soccer game, and organize RoboCup: The Robot World Cup Soccer Games and Conferences. In order for a robot team to actually perform a soccer game, various technologies must be incorporated including: design principles of autonomous agents, multi-agent collaboration, strategy acquisition, real-time reasoning, robotics, and sensor-fusion. RoboCup is a task for a team of multiple fast-moving robots under a dynamic environment. RoboCup also offers a software platform for research on the software aspects of RoboCup [2].

Football is more close to real life problems as chess, since it combines different skills for acting in a highly dynamic world which is only incompletely observable.

It took 50 years to beat the human champion in chess, more than 50 years may be necessary to develop robots which can play football in a reasonable way. A comparison between football and chess results in the characteristics given by Figure 1.

The first Robot World Cup Soccer Games "RoboCup 97" took place in 1997 in Nagoya (Japan) in connection with the International Joint Conference on Artificial Intelligence, IJCAI'97. The second Robot World Cup Soccer Games "RoboCup 98"

Chess	Football
static environment	rapidly changing situations
complete information	incomplete information
reliable information	unreliable information
3 minutes per move	quick reactions
	(RoboCup simulation league: 100ms)
single actions	action sequences, cooperation
single actor	team
abstract thought	observation - decision - action

Figure 1: Comparison between football and chess

took place in July 1998 in Paris, in parallel to the human World Cup. Competitions were held in 3 classes:

1. Middle size league: Robots with max. 50 cm ⊘.

2. Small size league: Robots with max. 15 cm ⊘.

3. Simulation league with matches in a virtual world.

The virtual soccer environment of the simulation league provides a very useful test-bed for various problems of cooperating autonomous systems acting in a very dynamic environment with incomplete and unreliable information. The material world (bodies of the players, ball, field, referee decisions) is simulated by the Soccer Server program, and visualized by the Soccer Monitor program which are available from the Web. The programs of the competitors implement the brains of the virtual players (cf. Section 2).

Our teams "AT Humboldt" were champion in the simulation league in 1997 and vice champion in 1998. They were implemented using agent-oriented techniques based on the mental concepts of BDI = belief – desire – intention (cf. Section 3).

Because of the huge amount of data and the continuum of different behaviors, direct programming of soccer players becomes more and more impossible: It has to be replaced by related learning techniques (cf. Section 4). This holds for the simulation league as well as for the real robots. Learning is possible at the different stages of:

- "atomic" actions (e.g. kick),

- skills (e.g. goal shooting, dribbling),

- cooperation (e.g. double pass),

- strategy (e.g. adaptation to opponents).

Reinforcement learning was applied for learning of skills by different teams, cf. e.g. [3]. Learning at different stages ("Layered Learning") is discussed in [4].

2   The Environment of the Simulation League

The simulator program of the soccer simulation for the virtual RoboCup consists of two subsystems, the *SoccerServer* and the *SoccerMonitor*. The SoccerServer is the simulator's engine, whereas the SoccerMonitor visualizes the game. Both programs are available from the net [5]. They are continuously improved to become more and more realistic as far as possible with respect to technical development.

The SoccerServer runs as a separate process and waits for the login of the 22 different player-agents (clients). From this point of view the SoccerMonitor is nothing else but a further client. The communication between the server and his clients runs via UDP; thus real distribution of the client-programs over several computers is possible. The player programs are not allowed to communicate directly with each other (only by using the say-command of the server).

When all the player–agents have logged in properly, the human observer can start the match by the SoccerMonitor and the SoccerServer starts simulation. Two interfaces are used in the communication between server and clients: The server sends (textual) strings coding visual and audible information in predefined intervals (typically each 150 msec) to the player programs. The players can react with transmitting (textual) commands from a given set of atomic actions.

The "visual" information of a player depends on its facing direction and the distance of observable objects. He can typically observe only objects in a region of 90 degree. Far players cannot be discriminated according to team identity or number, respectively. The player gets information about the distance of objects and their deviation from facing direction. Expected changes of moving objects are given for near objects. All information is subject to certain noise.

Players do not get information about their coordinates, but they get information about some landmarks (goals, lines, flags) as far as these objects are in their visible region.

"Auditorial" information from the referee is available for all players. The built-in referee of the SoccerServer actually judges goals, outside, offside and corner kick (the properties of the SoccerServer are subject to continuous development for better simulation). "Auditorial" information from other players (command *Say*) is available for all players in a distance of less than 50 m, and it is restricted to only one message per 200 msec.

Every agent can send commands to the SoccerServer in constant time slices of 100 ms. The agent is responsible for maintaining the time intervals (a second command in a 100 ms interval will not be considered by the server). Typical commands are *Kick*, *Dash*, *Turn* for acting on the field, and *Say* for "broadcasted" messages. The parameters of actions like power and direction are given relative to the facing direction of the agent.

An example of the communication between a player and the Soccer Server is given in Figure 2:

Visual information (per 150 *msec):*
                    ((flag 1 t) 105 19)
                    ((player) 76 -23)
                    ((player red) 36 -42)
                    ((player red 2) 4 12 0.23 -7)
                    ((ball) 1.8 14 -0.5 34)

Actions of players (only one action per 100 *msec):*
                    (turn 129)
                    (dash 100)
                    (kick 100, -129)

Figure 2: Examples of communication between a player and the Soccer Server by character strings.

The player can see the flag on the top of the left side in a distance of 105 m with·an angle of 19 degree right from its facing direction. The player with the distance of 76 m is seen left from its facing direction (-23 degree), it is too far to be identified. The player with the distance of 36 m is near enough to identify its team name "red". For near players like the one with distance 4 m, the player number "2" is observable. Moreover, the information for this player yields additional information of actual movement: After the next cycle of 100 ms, the player will change the distance by 0.23 m and the direction by -7 degree. Related information is available for the ball.

In this situation, the ball can be intercepted by the player. The player may therefore decide for the following plan consisting of 3 atomic actions which can be sent to the Soccer server with delays of 100 ms. It first turns for 129 degree, makes a step with speed "100" (that means maximal speed of 10 m/s; – to allow scoring during the short matches of 10 minutes, the parameters of the Soccer Server are set to allow higher speeds than in reality) and then kicks with power "100" into the direction of angle -129 relative to its facing direction.

Complex actions like shooting (stop the ball, place the ball, kick the ball) can be combined from related atomic actions. Consecutive kicks accelerate the ball.

Players have only restricted resources. They have a stamina which is decreased according to the power of dash commands. It is exhausted after a full speed run over about half of the field, after that it restored only slowly.

3   Architecture of AT Humboldt programs

3.1   Agent Oriented Programming

The consideration of programs as agents focuses at first on the aspect of autonomy: Programs have to act in an appropriate way to changes in the environment. Therefore they have some input or sensor facilities and some output or actoric components. The mapping from input to output can be done in very simple ways (e.g. strictly reactive) or in more sophisticated ways up to models which are inspired by human decision processes. We found that mental notions like capabilities/skills, belief, goals/desires and intentions/plans are very useful pictures to make agent programming transparent. The aspect of rationality forces agents to deal efficiently with their resources, especially with time.

Many different architecture have been proposed for the design of agents. The subsumption architecture [6] for example is a well known architecture with different levels of abstraction. The so-called BDI model fits best to our concept of agents in Artificial Soccer [1]. BDI stands for Belief-Desire-Intention, and the approach is based on the philosophical work of Bratman [7], and the theoretical and practical work of Rao and Georgeff [8, 9] and others (e.g. [10, 11]).

In the BDI-approach, agents maintain a model of their world which is called belief (because it might not be true knowledge). The way from belief to actions is guided by the desires of the agent. Bratman has argued that intentions are neither desires nor beliefs, but an additional independent mental category. Intentions are considered as (partial) plans for the achievement of goals by appropriate actions. Commitment to intentions is needed which has impact on the rational usage of resources: The (relative) stability of committed intentions prevents overload in deliberation and useless plan changes, and it serves trustworthiness in cooperation.

Traditionally, BDI approaches are embedded in frameworks of multi-modal temporal logics. This makes them unattractive for programming in procedural or object-oriented environments. We found that object-oriented programming is best suitable for programming our soccer agents, and we have chosen C++ for performance reasons. But the BDI approach gives us a good structure for our agent architecture. We will come back to this point at the end of this section.

3.2   Components of AT Humboldt

The agents of the AT Humboldt teams [12, 13, 14] can use different *skills* like GO-TO-POSITION, KICK, DRIBBLING, etc. . Each skill can be considered as a partial plan or a parameterized procedure. It consists of a sequence of atomic actions which are specified in detail according to the underlying situation.

The *belief* component models the belief of the agent about the environment according to sensor information and simulation for moving objects. It provides simulation and evaluation of speculative future developments for possible actions.

The *desire* component evaluates potential desires according to the beliefs. Typical desires are "intercept ball", "goal-kick", "dribbling". The commitment for a desire is based on the expected success by the simulation tools of the belief component.

The *intention* component specifies the best plan according to a committed desire. It uses the predefined skills. The intention component parameterizes these plans according to the expectation of optimal success (again on the base of simulation).

After deliberation is complete, a sequence of atomic actions is given to the *execution* component which is responsible for the synchronization with the Soccer Server.

The decision processes depend on the role of the players. The roles define special preferences which concern the direction of kicks and the frequency of dribbling, respectively. This emerges in special forms of cooperative play which gives preferences to a more aggressive play over the wingers.

3.3   Selection of Intentions

The agent can choose between different options (e.g. dribbling or shooting). From a theoretical view point, intentions have to be chosen according to the maximal expected utility, which has to be computed according to different *options o* which are realizable by different *plans p*, where we may have different possible *results r*. If $u(r)$ is the *utility* of a result r, then the expected utility of a plan p is computed by

$$u(p) = \sum_{r \text{ result of } p} \pi(r) \cdot u(r)$$

with related probabilities π. Then the utility of option o is given by

$$u(o) = Max\{ \ u(p) \mid p \text{ plan for } o \ \}.$$

In practice, these calculations are not tractable. Hence the decision process is performed as follows:

1. Option utilities $u(o)$ are approximated.

2. Best scoring options o are chosen as "desires".

3. More sophisticated evaluation is performed for desires.

4. Best scoring desire is chosen as "intention".

5. Parameters of a plan for the intention are computed.

A special problem is the trade-off between adaptation and stability of plans: "Fanatism" may follow a plan even if this plan is not useful anymore, while frequent adaptation may lead to endless changes of a plan. In our agents, new plans are calculated as often as new information is available. In many cases, the new plan is closely related to the old one which is still under execution. Small changes in the new plan are dropped. If the new plan is more different from the old one, then a special decision procedure is invoked. In this way, the "stability of intentions" [7] is maintained.

We finish this section with some remarks concerning agent-oriented software techniques. A new view to programs and machines stands behind the broad usage of the term "agent" – by both the insiders and the public. The consideration as autonomously acting entities is the main aspect, while other aspects like intelligence, cooperation, mobility etc. may vary from one application to another.

In our approach, procedures/methods are considered as the basic behavior of agents as well as of traditional programs. Actually, the skills of our agents are procedures. But the call of procedures in traditional programs follows a control flow which is strictly chosen already by the programmer. This is not possible in highly dynamic situations: The programmer can not explicitly define what the agent has to do in all possible situations. The agent has to analyze the situation and to decide autonomously by related criteria what to do. The programmer can implement only the methods for such decisions (to give an example: the programmer of a chess program does not know which move his program will do in a concrete situation). We found, that the BDI framework provides a very good structure for programming control mechanisms for agents.

4 Learning in RoboCup

We distinguish between off-line learning ("training") and on-line learning (adaptation during the matches).

The success of skills depends on appropriate choices of the consecutive actions. Thereby, learning should not concern a single action, but the whole sequence of actions. Learning for skills can be performed as off-line learning.

Choices in the deliberator depend on several parameters (especially for utility calculations). These parameters may be tuned in a general way (off-line learning), or regarding the behavior of an opponents during a match (on-line learning), respectively.

4.1 Off-Line Learning

As mentioned above, a shot is a combination of several atomic kick-actions. Each kick-action has its own parameters (power, direction) which determine the result together with the previous movement of the ball. The Soccer Server permits different amounts of the kick-power depending on the position of the ball. This makes it difficult to optimize a sequence of kicks by computation.

We are experimenting with the training of the ball-shooting skill. Therefore we collect data using different parameters in different situations in an automatized coach mode. This data is analyzed in order to find optimal parameters and hopefully to find computational rules for determining the optimal parameters according to different situations. Analysis can be done by special programs ("coach") outside of the player programs. After having some results, the next phase of training is to be done with improved parameters for further fine tuning.

Another trainable skill is the interception of a moving ball. It is important in the deliberation process, which players can intercept the ball first (e.g. for the decision if the agent should run to the ball, or for the decision concerning a pass to a team mate, respectively). Related computations are necessary several times during the deliberation

.process. Because of ball decay, the optimal point for fast interception is difficult to compute. In the first implementation, we used simulation of ball and players, which was extremely time consuming resulting in too long deliberation times. Now we have replaced the simulation by a scheme for the approximation of the optimal interception point, which was obtained from a set of generated examples.

In general terms, we consider off-line learning as the improvement of action sequences which are rawly described in the beginning and more and more improved by experience (as in human training, where the general course of actions is taught "mentally", but the fine tuning is performed by doing).

Further skills which we want to optimize in this way are dribbling and fighting for the ball.

4.2   On-Line Learning

On-line learning is used to adapt certain decisions and (collective) behaviors to the expected behavior of opponents. The short duration of a match makes it difficult or even impossible to use inductive learning (because of a lack of enough examples). Thus we base our on-line learning on techniques from Case Based Reasoning (CBR).

A good management of stamina is an important factor of success. This means to maintain good positions on the field and to maintain the control over the ball as long as possible. Good positions depend strongly on the opponents. We have experimented with records of the occurrences of the ball in a raw grid of the field. Then the players could adapt their positions to that knowledge. But the usage of this strategy did not improve the strength of our team. The reason for this result may be the simple reaction of opponent behavior according to the new position of our players.

In another experiment we have collected situations according to successful movements of players [15]. The neighborhood of the player was divided into segments, which were marked according to the players. In a new situation, the player retrieves similar cases from the case base and makes a move which is suggested by the previous situations. The decision process is illustrated by Figure 3: Two cases from the case base are shown in the upper half.
The left side of the lower half shows the present situation as "query". The query is extended by similarity activation to neighboring segments (lower half, right). In the retrieval process, the cases from the case base are evaluated according to their correspondence to the extended query: Case 1 is more similar, and hence the move direction will be chosen to the left.

The case base may be filled with cases from the present match as well as from other matches (especially of the same opponent).

Other applications of CBR may concern the behavior in standard situations and the treatment of the offside rule. In general, our understanding of CBR [16] means learning from former experiences (cases), especially for situations where we have not enough information to induce rules. Successful CBR needs efficient case memories

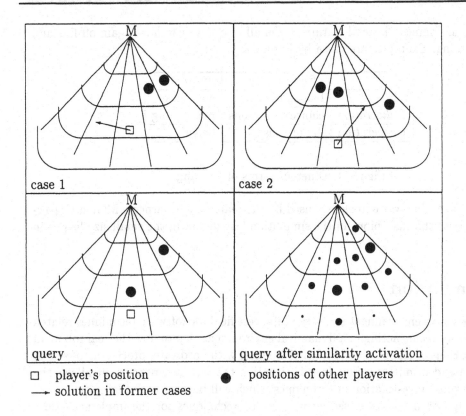

Figure 3: CBR for decisions concerning player movement.

which permit the retrieval ("reminding") of old cases in short time. In the example of Figure 3 we could manage the case retrieval from a case base with about 5000 cases in 30 ms [15].

RoboCup offers a lot of scenarios for learning from experiences. In off-line learning we can make experiments for collecting lots of cases, while in on-line learning we can collect only few cases in order to learn the opponents tactics and skills. Cases from off-line learning can be used to extract rules for behavior and to tune parameters (as we did for ball interception).

4.3   Layered Learning

Human football players have first to learn their individual skills before they can start with successful team play. In the same way, artificial players should first try to optimize their skills. Then they can learn how to use these skills successfully in different situations (if the skills are not optimal, they can only learn how to act with unreliable skills). Finally the agents can to learn cooperative and strategic play. Related learning approaches are named "layered learning". On all levels, we can have again off-line and on-line learning. Examples are given by Figure 4.

	off-line	*on-line*
Skills	ball-shooting	goal kick
Deliberation	choice of options	positioning
Cooperation	double pass	defense

Figure 4: Different forms of learning.

Layered Learning was extensively used by the team of the Carnegie Mellon University which were the champions in the simulation league and in the small size league in RoboCup98 [4].

5   Conclusion

The virtual soccer environment is a very useful test field for software techniques related to autonomous agents acting in a dynamical environment. Up to now the real robots in the competitions of RoboCup had a lot of troubles with hardware, processing of sensor information and primitive skills. In this situation the virtual environment allows the development and investigation of techniques which will become useful later on.

In our approach we have used agent oriented techniques for the implementation. We have performed some experiments with learning, especially in the field of CBR.

Acknowledgment

The design and implementation of the programs AT Humboldt were developed in courses for students at Humboldt University. The author want to give special thanks to Markus Hannebauer, Jan Wendler, Kay Schröter, Pascal Müller-Gugenberger and Helmut Myritz.

References

[1] Kitano, H., Kuniyoshi, Y., Noda, I., Asada, M., Matsubara, H. and Osawa, H.: RoboCup: A Challenge Problem for AI. AI Magazine 18(1): 73–85. 1997.

[2] Official RoboCup website: http://www.robocup.org

[3] Minoru Asada, Shoichi Noda, and Koh Hosoda: Action-Based Sensor Space Segmentation for Soccer Robot Learning. In: Applied Artificial Intelligence, Volume 12, Number 2-3, 1998.

[4] Peter Stone and Manuela Veloso: Layered Approach to Learning Client Behaviors in the RoboCup Soccer Server, In: Applied Artificial Intelligence, Volume 12, Number 2-3, 1998.

[5] Programs for Soccer Server and Soccer Monitor are available from the page http://ci.etl.go.jp/ noda/soccer/server/index.html

[6] Brooks, R.A., "Elephants don't play chess", In P.Maes, editor, Designing Autonomous Agents, MIT press, 1990.

[7] Bratman, M. E.: Intentions, Plans, and Practical Reason. Harvard University Press, 1987.

[8] Rao, A. S. and Georgeff, M. P.: An Abstract Architecture for Rational Agents. In Nebel, B., Rich, C. and Swartout, W. (eds.): Proc. of the Third International Conference on Principles of Knowledge Representation and Reasoning: 439–449. Morgan Kaufmann Publishers, 1992.

[9] Rao, A. S. and Georgeff, M. P.: BDI agents: From theory to practice. In Lesser, V. (ed.): Proc. of the First Int. Conf. on Multi-Agent Systems (ICMAS-95): 312–319. MIT-Press, 1995.

[10] Jennings, N. R.: Specification and Implementation of a Belief-Desire-Joint-Intention Architecture for Collaborative Problem Solving. Int. Journal of Intelligent and Cooperative Information Systems 2(3): 289–318. 1993.

[11] Wooldrige, M. J.: This is MyWorld: The Logic of an Agent-Oriented DAI Testbed. In Wooldrige, M. J. and Jennings, N. R. (eds.): Intelligent Agents: 160–178. LNAI 890. Springer, 1995.

[12] Burkhard, H.-D., Hannebauer, M. and Wendler, J.: Belief-Desire-Intention Deliberation in Artificial Soccer. AI Magazine 19(3): 87–93. 1998.

[13] Burkhard, H. D., Hannebauer ,M., Wendler, J. : AT Humboldt — Development, Practice and Theory. RoboCup-97: Robot Soccer World Cup I, Lecture Notes in Artificial Intelligence 1395, Springer 1998, 357–372.

[14] Müller-Gugenberger, P. and Wendler, J.: AT Humboldt 98 — Design, Implementierung und Evaluierung eines Multiagentensystems für den RoboCup-98 mittels einer BDI-Architektur. Diploma Thesis. Humboldt University Berlin, 1998.

[15] Jan Wendler and Mario Lenz: CBR for Dynamic Situation Assessment in an Agent-Oriented Setting. Proceedings of the AAAI-98 Workshop on Case-Based Reasoning Integrations (Eds.: D. Aha and J. Daniels), 1988,

[16] Lenz, M., Bartsch-Spörl, B., Burkhard, H. D., Wess, S.: Case Based Reasoning Technology. From Foundations to Applications. LNAI 1400, Springer 1998.

CONTROLLING BASED ON STOCHASTIC MODELS

H.-J. Lenz
Free University Berlin, Berlin, Germany

E. Rödel
Humboldt University Berlin, Berlin, Germany

0. INTRODUCTION

Operative Controlling is used in industry and commerce in order to find out whether there exist avoidable irregularities in a given data set or not. These irregularities may be caused by environment, management, and unhonest or erroneous actions of any kind. Controlling is model based in the sense that prior knowledge about facts and structural relationships between variables is used to build up a model of a firm. Its relations are based on definitions, institutional and behavioural equations.

It is reasonable to assume, that the variables in the model can not be metered or observed exactly, but that there exist errors in the variables. If errors in the data are allowed, almost always a given data set will violate the underlying structural relationships. Therefore, given a data set, i.e. a realisation of random variables considered, it is necessary to derive predictors and test the deviations between realised and predicted values for significance. This paper will focus on the prediction problem.

In the following, we consider linear as well as non-linear relations between variables, and limit ourselves to arithmetic operations. Typically, relations like profit = revenues - expenditures or total consumption = Σ consumptionrate$_i$ lead to linear relationships. Non-linear equations come in by relations like turnover = quantity*price or profitability = 100 * profit / sales. Linear structural relations have been investigated by Lenz and Rödel [1]. This approach will be extended in this paper by nonlinear relations.

1. THE MODEL

The first part of the model contains the vector ξ of variables observed with an additional vector of errors u ,.i. e.

$$X = \begin{pmatrix} x_1 \\ . \\ . \\ . \\ x_p \end{pmatrix} = \begin{pmatrix} \xi_1 \\ . \\ . \\ . \\ \xi_p \end{pmatrix} + \begin{pmatrix} u_1 \\ . \\ . \\ . \\ u_p \end{pmatrix} = \xi + u. \tag{1.1}$$

The second part describes the structural relations between the variables by a vector g of functions $g_i(\xi)$ (i=1,...,q) observed with an additional vector of errors v, too, i.e.

$$Z = \begin{pmatrix} z_1 \\ . \\ . \\ . \\ z_q \end{pmatrix} = \begin{pmatrix} g_1(\xi) \\ . \\ . \\ . \\ g_q(\xi) \end{pmatrix} + \begin{pmatrix} v_1 \\ . \\ . \\ . \\ v_q \end{pmatrix} = g(\xi) + v. \tag{1.2}$$

So we will use the following denotations :

X - vector of observed data,
ξ - vector of true data,
Z - vector of observed data transformations,
$g(\xi)$ - vector of true data transformations,
u,v - vectors of random errors,
$\Sigma_u := P$ -covariance matrix of u,
$\Sigma_v := R$ - covariance matrix of v,
$$\Sigma_{u,v} = \begin{pmatrix} P & 0 \\ 0 & R \end{pmatrix}, \quad \zeta = g(\xi).$$

We will assume

• $g(\xi)$ is twice continously differentiable,
• P and R are regular.

2. LEAST SQUARES ESTIMATION OF ξ AND $g(\xi)$

The least squares estimators (LSE) of ξ and $g(\xi)$ are defined by minimizing the sum of Mahalanobis distances between the observed data and the true data and between observed transformations und true transformations, respectively, i.e.

$$V = \|X - \xi\|^2_{P^{-1}} + \|Z - \zeta\|^2_{R^{-1}} \, , \quad \text{where} \quad \|Y\|^2_A = Y^T A Y, \tag{2.1}$$

and

$$\min_{\xi, \zeta} V$$

under $\zeta = g(\xi)$

leads to the equations

$$V^* = \|X - \xi\|^2_{P^{-1}} + \|Z - \zeta\|^2_{R^{-1}} + \lambda^T (g(\xi) - \zeta), \tag{2.2}$$

where $\lambda^T = (\lambda_1, \ldots, \lambda_q)$ is a vector of Lagrange multipliers. Equating the partial derivatives to zero we get

$$(1) \frac{\partial V^*}{\partial \xi} = -2 P^{-1}(X - \xi) + \sum_{i=1}^{q} \lambda_i \nabla g_i(\xi) = 0,$$

$$(2) \frac{\partial V^*}{\partial \zeta} = -2 R^{-1}(Z - \zeta) - \lambda = 0, \tag{2.3}$$

$$(3) \frac{\partial V^*}{\partial \lambda} = g(\xi) - \zeta = 0.$$

It holds by (2)
$$\lambda = -2 R^{-1}(Z - \zeta).$$

Substituting λ into (1) we have to solve the system of p nonlinear equations

$$\Psi(\hat{\xi}, X, Z) := -\frac{1}{2} \frac{\partial V^*}{\partial \xi}\Big|_{\xi = \hat{\xi}} = P^{-1}(X - \hat{\xi}) + G_{\hat{\xi}}^T R^{-1}(Z - g(\hat{\xi})) = 0, \tag{2.4}$$

where

$$G_\xi = \left(\frac{\partial g_i(\xi)}{\partial \xi_j} \right)_{\substack{i=1,\ldots,q \\ j=1,\ldots,p}}$$

is the Jacobian matrix of $g(\xi) = (g_1(\xi), \ldots g_q(\xi))^T$.

The identity

$$\Psi(\xi,\xi,g(\xi)) = 0 \qquad\qquad (2.5)$$

is true and it follows by well-known rules of matrix differentiation

$$\frac{\partial \Psi(\hat{\xi}, X, Z)}{\partial \hat{\xi}} = -P^{-1} + \frac{\partial}{\partial \hat{\xi}}(G_{\hat{\xi}}^{T} R^{-1}(Z - g(\hat{\xi})))$$

$$= -(P^{-1} + G_{\hat{\xi}}^{T} R^{-1} G_{\hat{\xi}}) + (Z - g(\hat{\xi}))^{T} R^{-1} * H_{g}, \quad (2.6)$$

where

$$H_{g} = \left(H_{g_{1}}, \quad \ldots, \quad H_{g_{q}}\right)^{T}$$

is stacked by the Hessians

$$H_{i} = \left(\frac{\partial^{2} g_{i}}{\partial \hat{\xi}_{j} \partial \hat{\xi}_{k}}\right)_{j,k=1,\ldots,p}$$

of the functions g_{i} and the operation $*$ is defined by

$$a^{T} * H_{g} := \sum_{i=1}^{q} a_{i} H_{i}, a^{T} = \left(a_{1}, \quad \ldots, \quad a_{q}\right).$$

The symmetric matrix

$$J_{\hat{\xi}} := \frac{\partial \Psi(\hat{\xi}, X, Z)}{\partial \hat{\xi}} \Big|_{\substack{\hat{\xi}=\xi \\ X=\xi \\ Z=g(\xi)}} = -(P^{-1} + G_{\xi}^{T} R^{-1} G_{\xi})$$

is regular and consequently, because of (2.5) and the Implicit Function Theorem there is a neighbourhood of $(\xi, \xi, g(\xi))$ in which a unique solution

$$\hat{\xi} = \psi(X, Z)$$

of (2.4) exists.

3. FIRST ORDER APPROXIMATIONS

Expanding

$$\hat{\xi} = \psi(X,Z)$$

in a Taylor series we get

$$\hat{\xi} = \psi(X,Z) \approx \psi(\xi,\zeta) + \frac{\partial \psi}{\partial X}\Big|_{\substack{X=\xi \\ Z=\zeta}}(X-\xi) + \frac{\partial \psi}{\partial Z}\Big|_{\substack{X=\xi \\ Z=\zeta}}(Z-\zeta)$$

and consequently

$$\hat{\xi} - \xi \approx A(X-\xi) + B(Z-\zeta)$$

since $\psi(\xi,\zeta) = \xi$ by (2.5) and because of the Implicit FunctionTheorem. For the same reason we have

$$A := \frac{\partial \psi}{\partial X}\Big|_{\substack{X=\xi \\ Z=\zeta}} = -\left(\frac{\partial \Psi}{\partial \hat{\xi}}\right)^{-1}\frac{\partial \Psi}{\partial X}\Big|_{\substack{\hat{\xi}=\xi \\ X=\xi \\ Z=\zeta}} = -J_\xi^{-1}P^{-1}$$

and

$$B := \frac{\partial \psi}{\partial Z}\Big|_{\substack{X=\xi \\ Z=\zeta}} = -\left(\frac{\partial \Psi}{\partial \hat{\xi}}\right)^{-1}\frac{\partial \Psi}{\partial Z}\Big|_{\substack{\hat{\xi}=\xi \\ X=\xi \\ Z=\zeta}} = -J_\xi^{-1}G_\xi^T R^{-1}.$$

The corresponding approximation for the covariance matrix of $\hat{\xi}$ is therefore

$$\Sigma_{\hat{\xi}} := APA^T + BRB^T = J_\xi^{-1}P^{-1}PP^{-1}J_\xi^{-1} + J_\xi^{-1}G_\xi^T R^{-1}RR^{-1}G_\xi J_\xi^{-1}$$

$$= (P^{-1} + G_\xi^T R^{-1}G_\xi)^{-1}(P^{-1} + G_\xi^T R^{-1}G_\xi)(P^{-1} + G_\xi^T R^{-1}G_\xi)^{-1}$$

$$= -J_\xi^{-1}.$$

and we have $\qquad \hat{\xi} - \xi \approx N(0,-J_\xi^{-1})$ $\hfill$ (3.1)

by the independence of X and Z and assuming that X and Z are normally distributed. Obviously , the inequality

$$\Sigma_{\hat{\xi}}^{-1} = -J_\xi \geq P^{-1}$$

is true and consequently

$$\Sigma_{\hat{\xi}} \leq P.$$

The least square estimator $\hat{\zeta}$ of ζ is in consequence of equation (3) of (2.3)

$$\hat{\zeta} = g(\hat{\xi}).$$

We approximate its distribution again by a Taylor series expansion.

$$\hat{\zeta} = g(\hat{\xi}) \approx g(\xi) + G_\xi(\hat{\xi} - \xi),$$

i. e.

$$\hat{\zeta} - \zeta \approx G_\xi(\hat{\xi} - \xi) \approx N(0, G_\xi \Sigma_\xi G_\xi^T). \qquad (3.2)$$

Replacing $\Sigma_{\hat{\xi}} = \Sigma_{\hat{\xi}}(\xi)$ by $\hat{\Sigma}_{\hat{\xi}} := \Sigma_{\hat{\xi}}(\hat{\xi})$ and G_ξ by $\hat{G}_\xi := G_{\hat{\xi}}$ we can approximate
σ-intervals , confidence regions and outlier tests.

If all the functions g_i (i=1,...,q) are linear we are completely in accordance with the
restricted linear regression model (cf [3],[4]).

4. EXAMPLES

4.1. Example 1
Let p=2, q=1. We use the interpretation ξ_1 = quantity-sold, ξ_2 = price per unit and

$\zeta = \xi_1 \xi_2$ as sales.

$$g(\xi_1, \xi_2) = \xi_1 \xi_2, \quad P = diag(\sigma_1^2, \sigma_2^2), R = \tau^2 I_q, \theta_1 := \frac{\sigma_1^2}{\tau^2}, \theta_2 := \frac{\sigma_2^2}{\tau^2}$$

(a) $\quad \Psi_1(\hat{\xi}, X, Z) = \sigma_1^{-2}(x_1 - \hat{\xi}_1) + \tau^{-2}(Z - \hat{\xi}_1 \hat{\xi}_2)\hat{\xi}_2 = 0$

(b) $\quad \Psi_2(\hat{\xi}, C, Z) = \sigma_2^{-2}(x_2 - \hat{\xi}_2) + \tau^{-2}(Z - \hat{\xi}_1 \hat{\xi}_2)\hat{\xi}_1 = 0$

$\Rightarrow \hat{\xi}_2 = (x_2 + \theta_2 Z \hat{\xi}_1) / (1 + \theta_2 \hat{\xi}_1^2)$

We get by substituting of ξ_2 into (a) and by simplification

$$P_1(\hat{\xi}_1, x_1, x_2, z) := -\theta_2^2 \hat{\xi}_1^5 + \theta_2^2 x_1 \hat{\xi}_1^4 - 2\theta_2 \hat{\xi}_1^3 + \theta_2(2x_1 - \theta_1 x_2 z)\hat{\xi}_1^2 + (\theta_1 \theta_2 z^2 - \theta_1 x_2^2 - 1)\hat{\xi}_1 +$$
$$x_1 + \theta_1 x_2 z = 0 \qquad (z := Z = (z_1))$$

for the LSE $\hat{\xi}_1$ of ξ_1.

Furthermore we have the following relations:

$$-\dot{J}_\xi = \begin{pmatrix} \sigma_1^{-2} + \tau^{-2}\xi_2^2 & \tau^{-2}\xi_1\xi_2 \\ \tau^{-2}\xi_1\xi_2 & \sigma_2^{-2} + \tau^{-2}\xi_1^2 \end{pmatrix},$$

$$\Sigma_{\hat\xi} = -J_\xi^{-1} = \frac{1}{1+\theta_2\xi_1^2 + \theta_1\xi_2^2}\begin{pmatrix} \sigma_1^2(1+\theta_2\xi_1^2) & -\sigma_1^2\theta_2\xi_1\xi_2 \\ -\sigma_1^2\theta_2\xi_1\xi_2 & \sigma_2^2(1+\theta_1\xi_2^2) \end{pmatrix}.$$

Summarizing we have by (3.1) and (3.2) the result

$$\hat\xi - \xi = \begin{pmatrix} \hat\xi_1 - \xi_1 \\ \hat\xi_2 - \xi_2 \end{pmatrix} \approx N(0, \Sigma_{\hat\xi}).$$

$$\hat\zeta - \zeta \approx N(0, \sigma_{\hat\zeta}^2),$$

where

$$\sigma_{\hat\zeta}^2 = \frac{\sigma_1^2\xi_2^2 + \sigma_2^2\xi_1^2}{1+\theta_2\xi_1^2 + \theta_1\xi_2^2}.$$

A numeric example

$$X = \begin{pmatrix} 10 \\ 20 \end{pmatrix}, Z = 150, \sigma_1 = 1, \sigma_2 = 2, \tau = 10.$$

Results : $\hat\xi_1 \approx 8.846$, $\hat\xi_2 \approx 17.693$, $\hat\zeta \approx 156.52$
1σ-intervals : 8.846 ± 0.754, 17.693 ± 1.508, 156.52 ± 9.286.

4.2 Example 2

Let p=2, q=1. We use the interpretation ξ_1 = profit, ξ_2 = sales and $\zeta = \xi_1/\xi_2$ as profitability of sales.

$$g(\xi_1,\xi_2) = \frac{\xi_1}{\xi_2} = \zeta, \xi_2 \geq c > 0$$

We assume again the covariance structure of example 1.

a) $\Psi_1(\hat{\xi},X,Z) = \sigma_1^{-2}(x_1 - \hat{\xi}_1) + \tau^{-2}(z - \frac{\hat{\xi}_1}{\hat{\xi}_2})\hat{\xi}_2^{-1} = 0,$

b) $\Psi_2(\hat{\xi},X,Z) = \sigma_2^{-2}(x_2 - \hat{\xi}_2) - \tau^{-2}(z - \frac{\hat{\xi}_1}{\hat{\xi}_2})\hat{\xi}_2^{-2}\hat{\xi}_1 = 0.$

Substituting

$$\hat{\xi}_1 = (\hat{\xi}_2^2 x_1 + \theta_1 z \hat{\xi}_2)(\hat{\xi}_2^2 + \theta_1)^{-1}$$

calculated from a) into b) we get for $\hat{\xi}_2$ the equation

$$-\hat{\xi}_2^5 + x_2\hat{\xi}_2^4 - 2\theta_1\hat{\xi}_2^3 + (2\theta_1 x_2 - \theta_2 z x_1)\hat{\xi}_2^2 + (\theta_2 x_1^2 - \theta_1\theta_2 z^2 - \theta_1^2)\hat{\xi}_2 + \theta_1\theta_2 z x_1 + \theta_1^2 x_2 = 0.$$

The matrix J_ξ and its inverse are given by the following matrices.

$$-J_\xi = \begin{pmatrix} \sigma_1^{-2}(1+\dfrac{\theta_1}{\xi_2^2}) & -\dfrac{\sigma_1^{-2}\theta_1\xi_1}{\xi_2^3} \\[2ex] -\dfrac{\sigma_1^{-2}\theta_1\xi_1}{\xi_2^3} & \sigma_2^{-2}(1+\dfrac{\theta_2\xi_1^2}{\xi_2^4}) \end{pmatrix},$$

$$\Sigma_\xi = -J_\xi^{-1} = \cfrac{1}{1+\dfrac{\theta_1}{\xi_2^2}+\dfrac{\theta_2\xi_1^2}{\xi_2^4}} \begin{pmatrix} \sigma_1^2(1+\dfrac{\theta_2\xi_1^2}{\xi_2^4}) & \dfrac{\sigma_1^2\theta_2\xi_1}{\xi_2^3} \\[2ex] \dfrac{\sigma_1^2\theta_2\xi_1}{\xi_2^3} & \sigma_2^2(1+\dfrac{\theta_1}{\xi_2^2}) \end{pmatrix},$$

$$\sigma_\zeta^2 = \frac{\sigma_1^2\xi_2^{-2} + \sigma_2^2\xi_1^2\xi_2^{-4}}{1+\theta_1\xi_2^{-2}+\theta_2\xi_1^2\xi_2^{-4}}.$$

A numeric example

$$X = \begin{pmatrix} 10 \\ 20 \end{pmatrix}, Z = 1, \sigma_1 = 1, \sigma_2 = 2, \tau = 0.1$$

Results : $\hat{\xi}_1 \approx 11.57, \hat{\xi}_2 \approx 15.21, \hat{\zeta} \approx 0.761$

1σ-intervals : $11.57 \pm 0.907, 15.21 \pm 1.535, 0.761 \pm 0.077$.

4.3 Example 3

Let p=2, q=2. The two variables considered are linked simultaneously together by a product and a quotient.

$$g_1(\xi_1, \xi_2) = \xi_1 \xi_2 = \zeta_1,$$

$$g_2(\xi_1, \xi_2) = \frac{\xi_1}{\xi_2} = \zeta_2, \xi_2 \geq c > 0$$

We assume the covariance structure

$$P = diag(\sigma_1^2, \sigma_2^2), R = diag(\tau_1^2, \tau_2^2).$$

a) $\Psi_1(\hat{\xi}, X, Z) = \sigma_1^{-2}(x_1 - \hat{\xi}_1) + \tau_1^{-2}(z_1 - \hat{\xi}_1 \hat{\xi}_2)\hat{\xi}_2 + \tau_2^{-2}(z_2 - \frac{\hat{\xi}_1}{\hat{\xi}_2})\hat{\xi}_2^{-1} = 0,$

b) $\Psi_2(\hat{\xi}, X, Z) = \sigma_2^{-2}(x_2 - \hat{\xi}_2) + \tau_1^{-2}(z_1 - \hat{\xi}_1 \hat{\xi}_2)\hat{\xi}_1 - \tau_2^{-2}(z_2 - \frac{\hat{\xi}_1}{\hat{\xi}_2})\hat{\xi}_2^{-2}\hat{\xi}_1 = 0.$

We could solve these equations in a similar way as in the examples 1 and 2 and would get an algebraic equation of the 10th degree. But it is more convenient to use the gradient method for solving these equations.
The matrix J_ξ is given by

$$\Sigma_{\hat{\xi}}^{-1} = -J_\xi = \begin{pmatrix} \sigma_1^{-2} + \tau_1^{-2}\xi_2^2 + \frac{\tau_2^{-2}}{\xi_2^2} & \tau_1^{-2}\xi_1\xi_2 - \frac{\tau_2^{-2}\xi_1}{\xi_2^3} \\ \tau_1^{-2}\xi_1\xi_2 - \frac{\tau_2^{-2}\xi_1}{\xi_2^3} & \sigma_2^{-2} + \tau_1^{-2}\xi_1^2 + \frac{\tau_2^{-2}\xi_1^2}{\xi_2^4} \end{pmatrix}.$$

A numeric example

$$X = \begin{pmatrix} 10 \\ 20 \end{pmatrix}, z_1 = 150, z_2 = 1, \sigma_1 = 1, \sigma_2 = 2, \tau_1 = 10.0, \tau_2 = 0.1.$$

Results : $\hat{\xi}_1 \approx 11.035, \hat{\xi}_2 \approx 13.884, \hat{\zeta}_1 \approx 153.206, \hat{\zeta}_2 = 0.795$,

1σ-intervals : 11.035 ± 0.603, 13.884 ± 0.836, 153.206 ± 9.255, 0.795 ± 0.077.

Furthermore we calculated the following correlations :

$$corr(\hat{\xi}_1, \hat{\xi}_2) = -0.450,$$

$$corr(\hat{\zeta}_1, \hat{\zeta}_2) = -0.107.$$

4.4 Summary of examples

Input-Table

Example -Nr.	1 Value	σ	2 Value	σ	3 Value	σ
x_1	10	1	10	1	10	1
x_2	20	2	20	2	20	2
z_1	150	10	-	-	150	10
z_2	-	-	1	0.1	1	0.1

Output-Table

Example -Nr.	1 Value	σ	2 Value	σ	3 Value	σ
$\hat{\xi}_1$	8.846	0.754	11.57	0.907	11.035	0.603
$\hat{\xi}_2$	17.693	1.508	15.21	1.535	13.884	0.836
$\hat{\zeta}_1$	156.52	9.286	-	-	153.206	9.255
$\hat{\zeta}_2$	-	-	0.761	0.077	0.795	0.077

The standard deviations of the transformations are almost constant, since the correlation between them is very weak.

5. NUMERICAL IMPLEMENTATION AND PARAMETER CONTROL

In all cases considered above the main difficulty consisted in the choice of appropriate start solutions for the newton method in Examples 1 and 2 and for the gradient method in example 3. It seems to be most favourable to take into account all first-look-informations given by the
observed data. We combined these informations by geometric averaging . For instance, in example 3 we recommend the start solution

$$\xi_1^{(0)} = sqrt(x_1 sqrt(z_1 z_2)), \; \xi_2^{(0)} = sqrt(x_2 sqrt(z_1 / z_2)).$$

By the choice of the gradient method for solving the system (2.4) we prefer robustness against unfavourable start solutions to a higher convergence speed (cf [2],[5]).
Another problem is caused by the assumed knowledge of the covariance matrices of X and Z, respectively. Because of the relation

$$\zeta = g(\xi)$$

we can assume in a first approximation under the condition that the inherent relations between the observations are not strongly violated

$$Z \approx g(X),$$

allthough X and Z are indepedently observed. Using this relation we can approximate Z and the standard deviations of the z_i-s by first order approximations

$$z_i - g_i(\xi) \approx \nabla^T g_i(\xi)(X - \xi) \quad (i=1,...,q)$$

and consequently assuming as in all examples the independence of the observations X_j (j=1,...,p) we have

$$\tau_i \approx \left(\sum_{j=1}^{p} \left(\frac{\partial g_i}{\partial \xi_j} \right)^2 \sigma_j^2 \right)^{1/2} \quad (i=1,...,q) \tag{5.1}$$

or still rougher

$$\tau_i \approx \tilde{\tau}_i := \left(\sum_{j=1}^{p} \left(\frac{\partial g_i}{\partial \xi_j} \right)^2_{|\xi=x} \sigma_j^2 \right)^{1/2}. \tag{5.2}$$

In example 3 we had

$$\tilde{\tau}_1 = \sqrt{800} \approx 28.3,$$

$$\tilde{\tau}_2 = \sqrt{2/400} \approx 0.071,$$

what may be in contradiction to the " given " values of $\tau_1 = 10$ and $\tau_2 = 0.1$. Generally we can expect that large differences between the $\tilde{\tau}_i$ and the τ_i (i=1,...,q) are caused by incorrect values of the variances of the z_i-s or by strong violations of the relations

$$\zeta = g(\xi).$$

But it is not clear how we can test differences between $\tilde{\tau}_i$ and τ_i . We will explain it by example 3.

Example 3 (continuation)

We have by (5.1) and the definition of the noncentral χ^2-distribution with ν degrees of freedom and noncentrality parameter δ, $\chi^2_\nu(\delta)$,

$$\tilde{\tau}_1^2 = \sigma_1^2 x_2^2 + \sigma_2^2 x_1^2 \doteq \sigma_1^2 \sigma_2^2 \chi_2^2 \left(\frac{\xi_1^2}{\sigma_1^2} + \frac{\xi_2^2}{\sigma_2^2} \right),$$

$$\tilde{\tau}_2^2 = \frac{\sigma_1^2}{x_2^2} + \frac{\sigma_2^2 x_1^2}{x_2^4} \doteq \frac{\sigma_1^2 \chi_2^2 \left(\frac{\xi_1^2}{\sigma_1^2} + \frac{\xi_2^2}{\sigma_2^2} \right)}{\sigma_2^2 \left(\chi_1^2 \left(\frac{\xi_2^2}{\sigma_2^2} \right) \right)^2}.$$

Tests for the variances τ_i based on the statistics $\tilde{\tau}_i$ depend on the unknown true data vector ξ . The investigation of this problem will be the topic of a later paper . It should be noted that the distribution of $\tilde{\tau}_2^2$ is not a standard distribution . It is clear that it will be so for the relations g in many applications, too.

6. REFERENCES

1. Lenz, H.-J. & Rödel ,E. : Statistical Quality Control of Data. Proceedings 16th Symposium on Operations Research(1991) ed. by R. Horst et al.
2. Jennings, A. & McKeown, J.J. : Matrix Computation. Wiley 1992.
3. Vinod, H.D. & Ullah, A. : Recent Advances in Regression Methods. Marcel Dekker 1981.
4. Toutenburg, H. & Rödel, E. : Mathematisch-statistische Methoden in der Ökonomie. Akademie-Verlag 1978.
5. Engeln-Müllges,G. & Reuter, F. : Formelsammlung zur numerischen Mathematik mit Turbo-Pascal-Programmen. Wissenschaftsverlag Mannheim/Wien/Zürich 1991

LIST OF PARTICIPANTS

Prof. Salem Benferhat
Institut de Recherche en Informatique de
Toulouse (I.R.I.T.)
Université Paul Sabatier * C.N.R.S.
F-31062 Toulouse Cedex 04, France
email: benferhat@irit.fr

Dr. Franz Fabris
Dept. of Mathematics and Informatics
University of Udine
via della Scienze, 206
I-33100 Udine, Italy
Tel.: +39 0432 55 8444
Fax: +39 0432 55 8499
email: fabris@dimi.uniud.it

Prof. Michael R. Berthold
Computer Science Division
University of California
329 Soda Hall
Berkeley, CA 94720 / USA
Tel.: +1 510 642 9827
Fax: +1 510 642 5775
e-mail: berthold@cs.berkeley.edu

Prof. Luigi Pace
Dept. of Statistics
University of Udine
v. Treppo, 18
I-33100 Udine, Italy
Phone: +39 0432 249570
Direct: +39 0432 249587
Fax: +39 0432 249595

Prof. Dr.sc. Hans-Dieter Burkhard
Inst. of Informatics
Humboldt University Berlin
D-10099 Berlin, Germany
Unter den Linden 6
Tel. +49 +30 2093 3167
Fax. +49 +30 20181 216
e-mail: hdb@informatik.hu-berlin.de

Prof. Dr. Günther Palm
Universität Ulm
Abteilung Neuroinformatik
Oberer Eselsberg
D-89069 Ulm, Germany
Tel.: +49 731 502 41 51
Fax: +49 731 502 41 56
e-mail: palm@neuro.informatik.uni-ulm.de

Prof. Giacomo Della Riccia
Dept. of Mathematics and Informatics
University of Udine
via della Scienze, 206
I-33100 Udine, Italy
Tel.: +39 0432 55 8419
Fax: +39 0432 55 8499
e-mail: dlrca@uniud.it

Prof. Adriano Pascoletti
Dept. of Mathematics and Informatics
University of Udine
via della Scienze, 206
I-33100 Udine, Italy
Tel.: +39 0432 55 8441
Fax: +39 0432 55 8499
email: pascolet@dimi.uniud.it

Prof. Moreno Falaschi
Dept. of Mathematics and Informatics
University of Udine
via della Scienze, 206
I-33100 Udine, Italy
Tel.: +39 0432 55 8472
Fax: +39 0432 55 8499
email: falaschi@dimi.uniud.it

Dr. Ing. Marino Petracco
illycaffe' s.p.a. - via Flavia, 110
34147 Trieste, Italy
Tel. +39 040 3890 397
email: petraccm@illy.it
http//: www.illy.com

Dr. Jörg Gebhardt
TU Braunschweig
Institut für Betriebssysteme und
Rechnerverbund
Bültenweg 74 - 75
D-38106 Braunschweig, Germany
Tel.: +49 531 391 3292
Fax: +49 531 391 5936
e-mail: gebhardt@ibr.cs.tu-bs.de

Prof. Alberto Policriti
Dept. of Mathematics and Informatics
University of Udine
via della Scienze, 206
I-33100 Udine, Italy
Tel.: +39 0432 55 8464
Fax: +39 0432 55 8499
email: policrit@dimi.uniud.it

Dr. Willi Klösgen
Gesellschaft für Mathematik und
Datenverarbeitung mbH
Postfach 12 40
D-53757 Sankt Augustin, Germany
Tel.: +49 2241 14 2723
Fax: +49 2241 14 2084
e-mail: kloesgen@gmd.de

Prof. Paolo Serafini
Dept. of Mathematics and Informatics
University of Udine
via della Scienze, 206
I-33100 Udine, Italy
Tel.: +39 0432 55 8442
Fax: +39 0432 55 8499
email: serafini@dimi.uniud.it

Prof. Dr. Rudolf Kruse
Otto-von-Guericke-Universität
Institut für Informatik
Universitätsplatz 2
D-39106 Magdeburg, Germany
Tel.: +49 391 67 18706
Fax: +49 391 67 12018
e-mail: rudolf.kruse@cs.uni-magdeburg.de

Dr. Arno Siebes
CWI
Kruislaan 413
NL-1098 SJ Amsterdam, The Netherlands
Tel.: + 31 20 592 41 39
Fax: +31 20 592 41 99
e-mail: arno@cwi.nl

Prof. Dr. Hans-Joachim Lenz
Freie Universität Berlin
Institut für Wirtschaftsinformatik
Garystr. 21
D-14195 Berlin, Germany
Tel.: +49 30 838 2380
Fax: +49 30 838 4051
e-mail: hjlenz@wiwiss.fu-berlin.de

Prof. Philippe Smets
Université Libre de Bruxelles
IRIDIA
50, avenue Franklin Roosevelt
C.P. 194/6
B-1050 Bruxelles, Belgium
secret: +32 2 650 27 29
GSM: +32 95 50 10 72
Tel.: +32 2 650 27 36
Fax: +32 2 650 27 15
e-mail: psmets@ulb.ac.be

Prof. Franco Malvestuto
Dipartimento di Scienze dell'Informazione
University of Rome (La Sapienza)
Via Salaria 113, I-OO198 Roma, Italy
Tel.: +39 649918310
fax : +39 68541842
e-mail: mal@dsi.uniroma1.it
Web page URL: http://www.dsi.uniroma1.it/

Prof. Dr. Robert Trappl
Austrian Research Institute
for Artificial Intelligence
Schottengasse 3
A-1010 Vienna, Austria
Phone: ++43-1-53532810
Fax: ++43-1-5320652
Email: robert@ai.univie.ac.at
Web: http://www.ai.univie.ac.at/oefai/

Prof. Angelo Montanari
Dept. of Mathematics and Informatics
University of Udine
via della Scienze, 206
I-33100 Udine, Italy
Tel.: +39 0432 55 8477
Fax: +39 0432 55 8499
email: montana@dimi.uniud.it

Dr. Paolo Vidoni
Dept. of Statistics
University of Udine
v. Treppo, 18
I-33100 Udine, Italy
Phone: +39 432 249570
Direct: +39 0432 249587
Fax: +39 0432 249595
e-mail: vidoni@dss.uniud.it

Printed in the United States
By Bookmasters